READING ESSE AND NOTE-TAKING GUIDE
STUDENT WORKBOOK

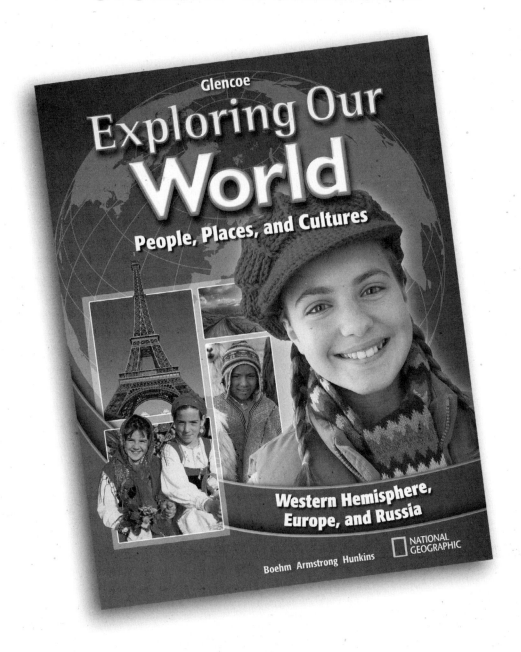

New York, New York   Columbus, Ohio   Chicago, Illinois   Woodland Hills, California

Glencoe

Copyright © by The McGraw-Hill Companies, Inc. All rights reserved. Except as permitted under the United States Copyright Act, no part of this publication may be reproduced or distributed in any form or by any means, or stored in a database or retrieval system, without prior permission of the publisher.

Send all inquiries to:
Glencoe/McGraw-Hill
8787 Orion Place
Columbus, OH 43240-4027

ISBN: 978-0-07-878170-4
MHID: 0-07-878170-1

Printed in the United States of America.

1 2 3 4 5 6 7 8 9 10 024 11 10 09 08 07

Contents

To The Student vii

Chapter 1 Using Geography Skills
Section 1 Thinking Like a Geographer 1
Section 2 The Earth in Space 4

Chapter 2 Earth's Physical Geography
Section 1 Forces Shaping the Earth 7
Section 2 Landforms and Water Resources 10
Section 3 Climate Regions 13
Section 4 Human-Environment Interaction 16

Chapter 3 Earth's Human and Cultural Geography
Section 1 World Population 19
Section 2 Global Cultures 22
Section 3 Resources, Technology, and World Trade 25

Chapter 4 Physical Geography of the United States and Canada
Section 1 Physical Features 28
Section 2 Climate Regions 31

Chapter 5 History and Cultures of the United States and Canada
Section 1 History and Governments 34
Section 2 Cultures and Lifestyles 37

Chapter 6 The United States and Canada Today
Section 1 Living in the United States and Canada Today 40
Section 2 Issues and Challenges 43

Chapter 7 Physical Geography of Latin America
Section 1 Physical Features 46
Section 2 Climate Regions 49

Chapter 8 History and Cultures of Latin America
Section 1 History and Governments 52
Section 2 Cultures and Lifestyles 55

Chapter 9 Latin America Today
Section 1 Mexico 58
Section 2 Central America and the Caribbean 61
Section 3 South America 64

Chapter 10 Physical Geography of Europe
Section 1 Physical Features 67
Section 2 Climate Regions 70

Contents

Chapter 11 History and Cultures of Europe
Section 1 History and Governments 73
Section 2 Cultures and Lifestyles 76

Chapter 12 Europe Today
Section 1 Northern Europe 79
Section 2 Europe's Heartland 82
Section 3 Southern Europe. 85
Section 4 Eastern Europe. 88

Chapter 13 Physical Geography of Russia
Section 1 Physical Features 91
Section 2 Climate and the Environment 94

Chapter 14 History and Cultures of Russia
Section 1 History and Governments 97
Section 2 Cultures and Lifestyles 100

Chapter 15 Russia Today
Section 1 A Changing Russia 103
Section 2 Issues and Challenges. 106

To The Student

Taking good notes helps you become more successful in school. Using this book helps you remember and understand what you read. You can use this *Reading Essentials and Note-Taking Guide* to improve your test scores. Some key parts of this booklet are described below.

The Importance of Graphic Organizers

First, many graphic organizers appear in this *Reading Essentials and Note-Taking Guide*. Graphic organizers allow you to see important information in a visual way. Graphic organizers also help you understand and summarize information, as well as remember the content.

The Cornell Note-Taking System

Second, you will see that the pages in the *Reading Essentials and Note-Taking Guide* are arranged in two columns. This two-column format is based on the **Cornell Note-Taking System,** developed at Cornell University. The large column on the right side of the page contains the essential information from each section of your textbook, *Exploring Our World.*

The column on the left side of the page includes a number of note-taking prompts. In this column, you will perform various activities that will help you focus on the important information in the lesson. You will use recognized reading strategies to improve your reading-for-information skills.

Vocabulary Development

Third, you will notice that vocabulary words are bolded throughout the *Reading Essentials and Note-Taking Guide*. Take special note of these words. You are more likely to be successful in school when you have vocabulary knowledge. When researchers study successful students, they find that as students acquire vocabulary knowledge, their ability to learn improves.

Writing Prompts and Note-Taking

Finally, a number of writing exercises are included in this *Reading Essentials and Note-Taking Guide.* You will see that many of the note-taking exercises ask you to practice the critical-thinking skills that good readers use. For example, good readers *make connections* between their lives and the text. They also *summarize* the information that is presented and *make inferences* or *draw conclusions* about the facts and ideas. At the end of each section, you will be asked to respond to two short-answer questions and one essay. The essays prompt you to use one of four writing styles: informative, descriptive, persuasive, or expository.

The information and strategies contained within the *Reading Essentials and Note-Taking Guide* will help you better understand the concepts and ideas discussed in your social studies class. They also will provide you with skills you can use throughout your life.

Chapter 1, Section 1 (Pages 14–17)

Thinking Like a Geographer

Big Idea

Geography is used to interpret the past, understand the present, and plan for the future. As you read, complete the chart below by identifying two examples for each topic.

Themes of Geography
1.
2.
Types of Geography
1.
2.
Geographers' Tools
1.
2.

 | **Read to Learn**

The Five Themes of Geography (page 15)

Explaining

Explain the difference between absolute location and relative location.

_____
_____
_____
_____
_____

Geography is the study of Earth and its people. Scientists who do this work are geographers. They use five main themes to describe people and places. The five themes of geography are location, place, human-environment interaction, movement, and regions.

The position of a place on Earth's surface is its *location,* which can be described in two ways. **Absolute location** refers to the exact spot on Earth where a place or feature is found. **Relative location** explains where a feature is in relation to the features around it.

Place refers to the characteristics of a location that make it unique. One way to define a place is by its physical features— landforms, plants, animals, and weather patterns. A place also can be defined by its human characteristics, such as its language.

The **environment** is one's natural surroundings. *Human-environment interaction* explores how people affect, and are affected by, their environment. People affect the environment by changing it to meet their needs. People, in turn, are influenced by environmental factors they cannot control, such as temperature and natural disasters.

The Five Themes of Geography (continued)

Applying

What region do you live in?

_____

Movement explores how and why people, ideas, and goods move from one place to another. For example, people might move to flee from a country that is at war. Movement causes cultural change.

Regions are areas of the Earth's surface that have features in common. These features may be land, natural resources, or population. For example, the Rocky Mountain region of the United States is known for ranching and mining.

A Geographer's Tools (pages 16–17)

Sequencing

Write down the four long periods of history from the earliest to the most recent.

1. _____
2. _____
3. _____
4. _____

Stating

What types of information do satellites provide to mapmakers?

_____
_____
_____

Types of Geography

Geographers study Earth's physical and human features. Physical geographers study land areas, bodies of water, plant life, and other physical features. They also examine natural resources and the ways people use them.

Human geographers study people and their activities. They examine religions, languages, and ways of life. Human geographers can focus on a specific location or look at broader areas. They often make comparisons between different places.

Places in Time

Geographers study history to learn about changes that have occurred over time. History is divided into blocks of time called periods. A **decade** is a period of 10 years. A **century** is a period of 100 years. A **millennium** is a period of 1,000 years.

In Western society, history is commonly grouped into four long periods. Prehistory is the time before writing was developed. This period ended about 5,500 years ago. The next period, which lasted until 1,500 years ago, is Ancient History. That period was followed by the Middle Ages, which lasted about 1,000 years. Modern History is the period from about 500 years ago through the present.

Map Systems

Geographers use maps to study different types of information about a place. Some maps are created from information collected by satellites that circle the Earth. For example, satellites provide photographs and can measure changing temperatures and pollution. A specific group of satellites makes up the **Global Positioning System (GPS).** This system uses radio signals to record the precise location of every place on Earth. GPS devices are installed in cars and trucks and used by hikers so they do not get lost.

A Geographer's Tools (continued)

Differentiating
What is the difference between GPS and GIS?

_____
_____
_____
_____

Geographic Information Systems (GIS) are computer hardware and software that collect geographic data and display the data on a screen. GIS provides more detailed information that does not usually appear on maps, such as types of soil and water quality.

Careers in Geography

Careers for geographers exist at all levels of government and in private businesses. Governments hire geographers to help determine how land and resources are best put to use. Geographers also study population trends and help plan cities. Businesses hire geographers to locate resources, decide where to set up new offices, and provide information about places and cultures that companies deal with.

Section Wrap-Up

Answer these questions to check your understanding of the entire section.

1. **Distinguishing** Explain the difference between *place* and *location*.

   _____

   _____

2. **Making Connections** Complete this chart with examples of what physical geographers and human geographers study.

Physical Geographers	Human Geographers

Think about the different career options for geographers. On a separate sheet of paper, write a paragraph about a job in geography that you might enjoy.

Chapter 1, Section 1

Chapter 1, Section 2 (Pages 34–38)
The Earth in Space

Big Idea

Physical processes shape Earth's surface. As you read, complete the diagram below. Explain the effects of latitude on Earth's temperature.

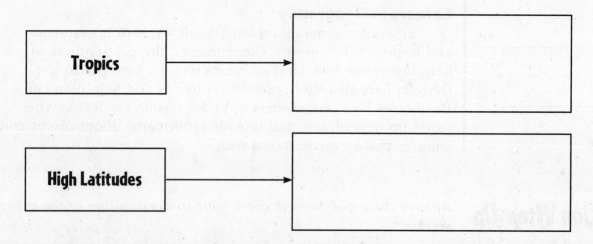

 **Notes** **Read to Learn**

The Solar System (pages 35–36)

Naming

Name the eight major planets in our solar system.

1. _____
2. _____
3. _____
4. _____
5. _____
6. _____
7. _____
8. _____

Eight major planets, including Earth, revolve around the sun. Thousands of smaller bodies also circle the sun. All of these, together with the sun, form our **solar system.**

Major Planets

The eight major planets differ from each other in size and form. The four inner planets closest to the sun are Mercury, Venus, Earth, and Mars. They are relatively small and solid.

Jupiter, Saturn, Uranus, and Neptune are the four outer planets. They are larger and formed mostly or entirely of gases. Pluto, once considered a major planet, is now classified as a minor planet.

Each planet follows its own **orbit,** or path, around the sun. Some orbits are almost circular, whereas others are oval shaped. The lengths of the orbits also vary, from 88 days for Mercury to 165 years for Neptune.

Earth's Movement

Earth makes a **revolution,** or complete circuit, around the sun every 365¼ days. This time period is defined as one year.

The Solar System (continued)

Explaining

Why do people not feel Earth move?

_____
_____

Every four years is a **leap year,** when the extra fourths of a day are combined and added to the calendar as February 29.

Earth **rotates,** or spins, on its axis as it orbits the sun. The **axis** is an imaginary line that passes through the center of Earth from the North Pole to the South Pole. Earth rotates in an easterly direction. It takes 24 hours for Earth to complete a single rotation. As it rotates, different parts of Earth are in sunlight, which is defined as daytime. Those parts facing away from the sun are in darkness and experience night. A layer of oxygen and gases, called the **atmosphere,** surrounds Earth. As Earth rotates, the atmosphere moves with it, so people do not feel Earth moving.

Sun and Seasons (pages 37–38)

Summarizing

What causes Earth to experience changing seasons?

_____
_____

Paraphrasing

Fill in the blanks. The day with the _____ *of sunlight is the beginning of summer. The first day of winter is the day with the* _____ *of sunlight.*

Earth is tilted 23½ degrees on its axis. This tilt causes seasons to change as Earth orbits the sun. The tilt determines whether or not an area will receive direct rays from the sun. When a hemisphere receives direct rays, it has summer. When a hemisphere receives indirect, or slanted, rays, it experiences the cold of winter.

Solstices and Equinoxes

The North Pole is tilted toward the sun on or about June 21, and the sun is directly over the Tropic of Cancer (23½°N latitude). This day is called the **summer solstice.** In the Northern Hemisphere, June 21 has the most hours of sunlight and marks the beginning of summer. On the same day, the Southern Hemisphere has the fewest hours of sunlight, and winter begins there.

Six months later, on or about December 22, the North Pole is tilted away from the sun and the sun's direct rays hit the Tropic of Capricorn (23½°S latitude). This is the **winter solstice** for the Northern Hemisphere—the day with the fewest hours of sunlight and the beginning of winter. It is the first day of summer in the Southern Hemisphere, however.

Midway between the two solstices are the **equinoxes,** when day and night are of equal length in both hemispheres. The equinoxes mark the beginning of spring and fall. The spring equinox occurs on or about March 21, and the fall equinox occurs around September 23. On both days, the noon sun shines directly over the Equator.

Chapter 1, Section 2

Notes | Read to Learn

Sun and Seasons (continued)

Identifying

Identify each latitude region.

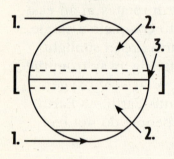

1. _____
2. _____
3. _____

Effects of Latitude

The **Tropics** is the low-latitude region near the Equator between the Tropic of Cancer and the Tropic of Capricorn. The sun's rays hit this area directly year-round, so temperatures in the Tropics tend to be warm. In contrast, the sun's rays are always indirect at the high-latitude areas near the North and South Poles. These polar regions are always cool or cold. The areas between the Tropics and the polar regions are called the midlatitudes. Temperatures, weather, and the seasons vary widely in these areas.

Section Wrap-Up

Answer these questions to check your understanding of the entire section.

1. **Explaining** How long is Earth's orbit? How long is Earth's rotation?

   _____

   _____

2. **Organizing** Complete this chart of seasons in the Northern Hemisphere. Add the approximate date when each season begins and the name of the first day for each season.

Season	Date Season Begins	Name of First Day
Winter		
Spring		
Summer		
Fall		

On a separate sheet of paper, explain why day and night are not always the same length throughout the year.

Chapter 2, Section 1 (Pages 44–48)
Forces Shaping the Earth

Big Idea

Physical processes shape the Earth's surface. As you read, complete this diagram by listing the forces shaping Earth and the effects of each.

Forces	Effects
_____	_____
_____	_____
_____	_____
_____	_____
_____	_____

 **Notes** | **Read to Learn**

Inside the Earth (page 45)

Labeling

As you read, write the name of each layer of Earth beside the correct number.

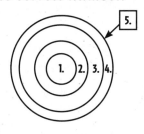

1. _____
2. _____
3. _____
4. _____
5. _____

The ground feels solid when you walk on it. But Earth is not a large, solid rock. Earth has several layers, like a melon or a baseball. The three main layers of the Earth are the core, the mantle, and the crust.

Scientists divide the **core** into the inner core and the outer core. At the center of the Earth is a solid inner core of iron and nickel. It is about 3,200 miles below the surface. Scientists think the inner core is under great pressure. The next layer, the outer core, is so hot that the metal has melted into a liquid.

The **mantle** surrounds the core. It is a layer of hot, thick rock. The section of the mantle nearest the core—the inner mantle—is solid. However, the rock in the outer mantle can be moved, shaped, and even melted. The melted rock is called **magma.** It flows to the surface of the Earth when a volcano erupts. Magma is called lava when it reaches the surface.

The thin, outside layer of the Earth is the **crust.** It is a rocky shell that forms the Earth's surface. The crust includes the ocean floors. It also includes seven large land areas known as **continents.** The continents are North America, South America, Europe, Asia, Africa, Australia, and Antarctica.

Chapter 2, Section 1

Read to Learn

Shaping the Earth's Surface (pages 46–48)

Identifying

As you read, identify two forces inside the Earth and two forces outside the Earth that change the appearance of Earth.

Inside the Earth:

_____

_____

Outside the Earth:

_____

_____

Explaining

How and why do the continents move?

_____

_____

_____

_____

_____

The Earth's crust moves and changes over time. Old mountains are worn down, and new mountains are pushed higher. The continents also move.

Plate Movements

The theory of **plate tectonics** explains how the continents were formed and why they move. Each continent sits on one or more large bases, called plates. As these plates move, the continents move. This movement is called continental drift. Earth's plates move very slowly, about 1 to 7 inches per year. Some 200 million years ago, all the continents were joined together in a huge landmass that scientists named Pangaea.

When Plates Meet

The movements of the Earth's plates shape the surface of the Earth. The plates may pull away from each other. This movement usually occurs in ocean areas, although it is also happening in East Africa. Plates also can run into each other, or collide. When two continental plates collide, they push against each other with tremendous force. The land where the plates meet is pushed up, which forms mountains.

Collisions of continental and oceanic plates produce a different result. The ocean plate is thinner. It slides underneath the continental plate. As the oceanic plate is forced down, magma in the Earth's mantle builds up. Volcanic mountains form as the magma erupts and hardens.

Earthquakes are sudden, violent movements in the Earth's crust. Many earthquakes happen in areas where the collision of oceanic plates and continental plates makes the Earth's crust unstable.

Earth's plates also can move alongside each other. This movement makes cracks in the Earth's crust called **faults.** Movements along faults result in sudden shifts that cause earthquakes.

Weathering

Volcanoes and earthquakes may cause immediate and drastic changes to Earth's surface. But other factors continue to change the landforms of the Earth at a slower pace. **Weathering** occurs when water, ice, chemicals, and even the roots of plants break rocks apart into smaller pieces.

Erosion

Weathered rock is then moved by water, wind, and ice in a process called **erosion.** Rivers and streams cut through mountains and hills. Ocean waves pound at rocks on the coast. Wind carries small bits of rock, which wear down larger rocks.

Chapter 2, Section 1

Section Wrap-Up

Answer these questions to check your understanding of the entire section.

1. **Identifying** What makes up the Earth's crust?

   _____
   _____
   _____

2. **Applying** What is plate tectonics, and what does it have to do with the Earth's shape?

   _____
   _____
   _____
   _____

Descriptive Writing

In the space provided, write an article for a science magazine describing how a valley was formed from weathering or erosion.

_____
_____
_____
_____
_____
_____
_____
_____
_____
_____
_____
_____

Chapter 2, Section 2 (Pages 49–54)

Landforms and Water Resources

Big Idea

Geographic factors influence where people settle. As you read, complete this diagram by identifying the various bodies of water that can be found on Earth's surface.

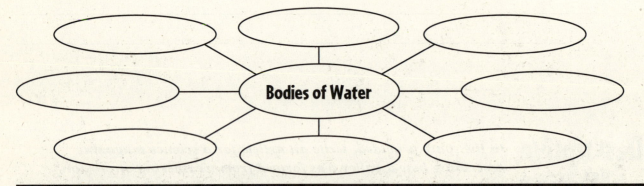

Notes | Read to Learn

Types of Landforms (pages 50–52)

Specifying

Which landforms are on continents, and which are on ocean floors?

On continents:

_____

_____

_____

On ocean floors:

_____

_____

Earth has many landforms, from mountains to lowlands. Some of the landforms can be found both on the continents and on the ocean floors.

On Land

Mountains—huge towers of rock—are the highest landforms. Hills are lower and more rounded than mountains. The long stretches of land found between mountains or hills are called valleys. There are two types of flatlands. Plains are flatlands in low-lying areas, frequently found along coasts and river valleys. Flatlands at higher elevations are called plateaus.

An isthmus is a narrow strip of land that connects two larger landmasses. It has water on two sides. A peninsula is a piece of land that is connected to a larger landmass on one side but has water on the other three sides. An island is a body of land smaller than a continent and is completely surrounded by water.

Under the Oceans

Off the coast of any continent is an underwater plateau called a **continental shelf**. Continental shelves stretch for

Types of Landforms (continued)

several miles until finally dropping sharply to the ocean floor. On the ocean floor, tall mountains rise along the edges of the ocean plates. Tectonic activity also makes deep cuts in the ocean floor called **trenches.**

Humans and Landforms

People live on all types of landforms. Climate—the average temperature and rainfall—is one factor that helps people decide where to live. The availability of resources is another factor.

The Water Planet (pages 52–54)

Categorizing

Name the bodies of water that are salt water and those that are freshwater.

Salt water:

_____

_____

_____

_____

Freshwater:

_____

_____

_____

Nearly 70 percent of Earth's surface is covered with water. Earth is covered with so much water that it is sometimes called the "water planet." Water exists in many forms. It can be found in liquid form, such as streams, rivers, lakes, seas, and oceans. The atmosphere holds water in the form of a gas known as vapor. Glaciers and ice sheets are made up of water that has been frozen solid.

Salt Water

All of the oceans on Earth are actually one large body of salt water. Most of the water on Earth—almost 96 percent—is salt water. Oceans flow into smaller areas—seas, bays, and gulfs—that are somewhat enclosed by land. These, too, hold salt water, and they are linked to oceans by narrow bodies of water called straits or channels.

Freshwater

Only 4 percent of the water on Earth is freshwater, and much of this is frozen in polar ice. **Groundwater** is also freshwater. It filters through the soil, often flowing through underground layers of rock called **aquifers.** Many communities get their freshwater by pumping it from aquifers.

Freshwater is also found in lakes and rivers. A lake is a large inland body of water. Most, but not all, lakes are freshwater. Rivers are long, flowing bodies of water that begin at a source and end at a mouth. The mouth is where a river empties into another body of water, such as an ocean or a lake.

Large rivers often have tributaries, or smaller rivers and streams that feed into them. Many rivers also form deltas at their mouths. A delta is an area where a river separates into streams that flow toward the sea. Soil carried by the river settles in the delta, eventually building up the land there.

Chapter 2, Section 2

Notes | Read to Learn

The Water Planet (continued)

Listing

List the four steps in the water cycle.

1. _____
2. _____
3. _____
4. _____

The Water Cycle

The total amount of water on Earth does not change, but it does move from place to place. Water circulates in a process called the **water cycle,** moving from the oceans, to the air, to the ground, and back to the oceans.

The sun's heat begins the water cycle by evaporating the water on the Earth's surface. **Evaporation** changes water from a liquid to water vapor. Water vapor rises from oceans and other bodies of water and spreads throughout the atmosphere.

When the air temperature cools, **condensation** takes place, meaning the water changes back into a liquid. Droplets of water in clouds fall to the ground as **precipitation.** This can take the form of rain, snow, sleet, or hail. Water then collects on the ground and in rivers, lakes, and oceans. This last step of the water cycle is known as **collection.**

Section Wrap-Up

Answer these questions to check your understanding of the entire section.

1. **Specifying** How does water help you determine whether a landform is an isthmus, a peninsula, or an island?

   _____
   _____

2. **Illustrating** Draw a diagram to show how the water cycle works. Be sure to label the various parts.

On a separate sheet of paper, write a paragraph explaining what you think the advantages and disadvantages are of living on a mountain or living in a valley.

Chapter 2, Section 3 (Pages 55–61)
Climate Regions

Big Idea

Geographers organize the Earth into regions that share common characteristics. As you read, complete this diagram by identifying the effects of El Niño and La Niña.

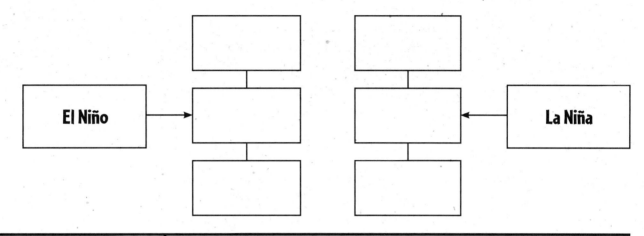

	**Read to Learn**

Nativism Resurges (page 610)

Listing

List three main things that affect climate.

1. _____
2. _____
3. _____

Geographers study weather and climate. **Weather** is the short-term changes in temperature, wind direction and speed, and air moisture in a particular location. **Climate,** in contrast, is the long-term, predictable patterns of weather for a region.

The Sun

The sun's heat directly affects Earth's climate. Earth is not heated evenly by the sun. The Tropics receive more of the sun's heat energy than do the Poles. The movement of air and water, however, helps spread the sun's heat around the globe.

Winds

Movements of air are called winds. Although winds can blow in any direction, major wind systems follow patterns called **prevailing winds.** For example, warm prevailing winds in the Tropics move north and south toward the Poles. Cold polar winds move toward the Equator.

Winds curve around the Earth as it rotates. Winds that blow from east to west between the Equator and the Tropics are called trade winds. Winds that move from west to east between the Tropics and 60° north latitude are known as the westerlies.

Notes | Read to Learn

Effects on Climate (continued)

Finding the Main Idea

Write the main idea of this subsection.

_____
_____
_____
_____
_____

Storms occur when moist, warm air rises suddenly and meets dry, cold air. Some storms include thunder and lightning as well as heavy rain. Storms may also include tornadoes—violent, funnel-shaped windstorms—or turn into blizzards.

Even more destructive storms are hurricanes and typhoons. These powerful storms begin in the Tropics and can grow to 300 miles across before striking coastal areas. Hurricanes occur in the western Atlantic and eastern Pacific Oceans. Hurricanes in the western Pacific Ocean are called typhoons.

Ocean Currents

Steadily flowing ocean streams are called **currents.** Like winds, they follow patterns and can affect climate. Ocean currents in the Tropics, for example, flow to higher latitudes and warm the winds that blow over them.

El Niño and La Niña

Two changes in Pacific wind and water patterns cause unusual weather every few years. Weakened winds allow warmer waters to reach the western coast of South America. This condition, called **El Niño,** results in heavy rains and flooding there and severe storms in North America. Yet little rain falls in Australia, southern Asia, and Africa.

La Niña, in contrast, brings low rainfall and cooler weather to the eastern Pacific. But in the western Pacific, La Niña produces storms with heavy rains, and typhoons may develop.

Landforms and Climate (pages 58–59)

Determining Cause and Effect

Why is the temperature on the tops of mountains very cold?

_____
_____
_____
_____
_____

The shape of the land, as well as its nearness to water, affects the climate. Some landforms cause **local winds,** which are wind patterns found in a small area. Some local winds form near water. These winds occur because land warms and cools more quickly than water.

Mountains, Temperature, and Rainfall

Mountains can affect the temperature and rainfall of a local area. The air is thinner at higher elevations. The temperature at the tops of mountains is often very cold because the thin air cannot hold heat well.

A **rain shadow** occurs when mountains block rain from reaching interior regions. The side of a mountain facing the wind—the windward side—can get a large amount of rainfall. The land on the other side of the mountain—the leeward side—can be very dry. Deserts can develop on the leeward side.

Read to Learn

Climate Zones (pages 59–61)

Listing

List the five major climate zones on Earth.

1. _____
2. _____
3. _____
4. _____
5. _____

A **climate zone** is an area with a particular pattern of temperature and precipitation. Areas in different parts of the world can have the same climate zone, meaning they have a similar climate. They also will have similar vegetation. Climate zones include **biomes**. These are areas in which certain kinds of plants and animals have adapted to the climate. Examples of biomes are rain forest, desert, grassland, and tundra.

Major Climates

Earth has five major climate zones—tropical, dry, midlatitude, high latitude, and highland. Four of these zones also have subcategories. For example, the dry climate zone is further divided into steppe and desert subcategories. These mostly dry climates vary slightly in rainfall and temperature.

Urban Climates

Large cities, or urban areas, often have a climate that is different from their surrounding areas. **Urban climates** have higher temperatures and unusual local winds. Paved streets and stone buildings soak up and release more of the sun's heat. The different heat patterns cause winds to blow into cities from several directions.

Section Wrap-Up

Answer these questions to check your understanding of the entire section.

1. **Determining Cause and Effect** How can winds affect the weather?
   _____
   _____

2. **Specifying** What three factors define a particular climate zone?
   _____
   _____

On a separate sheet of paper, write a description of the trade winds used by early explorers. Include in your account an additional description of the "horse latitudes" and why they were called that.

Chapter 2, Section 3

Chapter 2, Section 4 (Pages 63–66)

Human-Environment Interaction

Big Idea

All living things are dependent upon one another and their surroundings for survival. As you read, complete this chart by identifying four environmental problems and what people are doing to solve them.

Problem	Solution

Read to Learn

The Atmosphere (page 64)

Listing

As you read, list five negative effects of air pollution.

1. _____
2. _____
3. _____
4. _____
5. _____

People burn oil, coal, and gas for electricity, to power factories, and to move cars. Unfortunately, burning oil, coal, and gas causes air pollution.

Air Pollution

Air pollution can lead to smog. **Smog** is a thick haze of smoke and chemicals. People may have breathing problems when smog settles above cities.

Chemicals in air pollution can combine with precipitation to form **acid rain.** Acid rain kills fish, eats away at the surfaces of buildings, and destroys trees and other plant life.

Ozone is found in the atmosphere. It provides a shield against damaging rays from the sun. Some human-made chemicals destroy the ozone layer. This may result in more humans getting skin cancer.

The Greenhouse Effect

The **greenhouse effect** occurs when gases in the atmosphere trap the sun's heat. Because this heat is trapped, Earth remains warm, allowing living things to survive. Without the greenhouse effect, Earth would be too cold for most life-forms.

The Atmosphere (continued)

Differentiating

How is the greenhouse effect positive and negative?

_____
_____
_____

Some scientists believe air pollution strengthens the greenhouse effect. They claim that the burning of coal, oil, and natural gas traps the sun's heat near the Earth's surface and raises the planet's overall temperature. Such global warming could cause climate changes, melt polar ice, and result in flooding of coastal areas.

Many nations are trying to reduce global warming. Their focus is on using oil and coal more efficiently and cleanly. They also are looking at forms of energy that do not pollute. These include wind and solar power.

The Lithosphere (page 65)

Finding the Main Idea

Write down the main idea of the passage.

_____
_____
_____
_____
_____
_____

The Earth's crust is also known as the lithosphere. It includes all land above and below the oceans. Activities such as farming and cutting down trees can harm the lithosphere.

Topsoil is an important part of the lithosphere. It can be carried away by wind or water. Farming also puts topsoil at risk. However, farmers can limit the loss of topsoil in several ways. Instead of plowing straight rows, they can plow along the curves of the land. This is called contour plowing. Farmers also use **crop rotation,** which means they rotate or change what is planted from year to year. In addition, farmers plant grasses in fields without crops. The grasses hold the topsoil in place.

Another danger to topsoil is **deforestation,** or cutting down forests without replanting them. Tree roots hold soil in place. When trees are cut, the topsoil may be eroded. Trees are important for other reasons too. They support the water cycle and replace oxygen in the atmosphere. Forests are also home to many plants and animals.

The Hydrosphere and Biosphere (page 66)

Defining

Define the biosphere and the hydrosphere.

_____
_____

Surface water and groundwater make up the Earth's hydrosphere. Water is necessary for human life, but the amount of freshwater is limited. Therefore, people should practice **conservation.** This means resources such as water should be used carefully so they are not wasted.

Irrigation is a way farmers collect water and then use it on their crops. Irrigation often wastes water. Much of the water evaporates or soaks into the ground before it even gets to the crops.

Chapter 2, Section 4

Notes | Read to Learn

The Hydrosphere and Biosphere (continued)

Pollution harms the hydrosphere. **Pesticides** are chemicals that farmers use to kill insects. Strong pesticides and other chemicals sometimes spill into clean water. Then the water supply is threatened.

The biosphere is the "living" part of the planet—all the plants and animals. The biosphere is divided into many ecosystems. An **ecosystem** is a particular environment shared by certain plants and animals that need one another to survive.

Earth's **biodiversity,** or the variety of plants and animals, is getting smaller. Human activities have led to fewer types of plants and animals in some ecosystems.

Section Wrap-Up

Answer these questions to check your understanding of the entire section.

1. **Identifying** What human activities contribute to global warming?

   _____

   _____

2. **Determining Cause and Effect** How does contour plowing prevent the loss of topsoil?

   _____

   _____

Informative Writing

Create an outline for a presentation about the ways in which the hydrosphere is threatened.

_____

_____

_____

_____

_____

_____

Chapter 3, Section 1 (Pages 72–76)
World Population

Big Idea

The world's population is increasing, yet people live on only a small part of the Earth's surface. As you read, complete the diagram below to show the causes and effects of global migration.

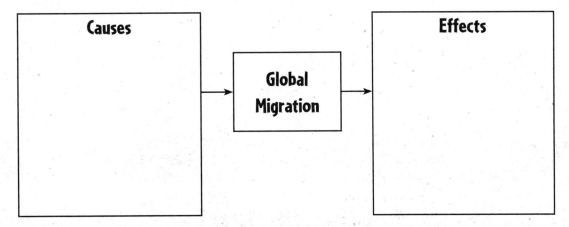

Notes | Read to Learn

Population Growth (page 73)

List two challenges created by rapid population growth.

1. _____
_____

2. _____
_____

The world's population has grown quickly in the past 200 years. One billion people lived on Earth around 1800. Today the population has risen to more than 6 billion. This rapid growth creates challenges for many countries.

One reason the population has grown so rapidly is because the death rate has decreased. The **death rate** is the number of deaths per year for every 1,000 people. The death rate has gone down because of better health care, improved living conditions, and an increase in the food supply.

A second cause of the population growth is high birthrates in Asia, Africa, and Latin America. The **birthrate** is the number of births per year for every 1,000 people.

Rapidly growing populations often experience food shortages. New **technology,** such as better irrigation methods and hardier crops, has helped increase the supply of food. But **famine,** or a severe lack of food, can occur as a result of warfare or crop failures.

Rapidly growing populations may experience shortages of clean water and housing as well. They also may lack hospital services and good schools.

Notes — Read to Learn

Where People Live (page 74)

Identifying
In what regions are most of the world's people clustered?

_____
_____
_____
_____

Only 30 percent of Earth is covered by land. Further, only half of this land can support large numbers of humans.

Population Distribution
People tend to live in areas that have fertile soil, mild climates, natural resources, and water. Therefore, the world's population is not evenly spread over the land. About two-thirds of the people on Earth live in just five regions—East Asia, South Asia, Southeast Asia, Europe, and eastern North America.

Population Density
The average number of people living in a square mile or square kilometer is called **population density.** Geographers use this measurement to figure out how crowded an area is. To determine the population density of an area, geographers divide the total population by the total land area.

Two countries can have the same amount of land but a different population density. For example, Norway and Malaysia both have about 130,000 square miles. But Norway's population density is 40 people per square mile, whereas Malaysia's population density is 205 people per square mile.

Population Movement (pages 75–76)

Defining
What is urbanization?

_____
_____
_____

Differentiating
How are refugees different from emigrants?

_____
_____
_____

For thousands of years, people have moved from one place to another. People and groups continue to move today.

Types of Migration
When people move from place to place within a country, it is called internal migration. These people often move from a farm or rural area into a city. Such movement leads to **urbanization,** or the growth of cities.

Movement from one country to another is called international migration. When people **emigrate,** they leave their birth country. People who emigrate are called *emigrants* in their home country and *immigrants* in their new country.

Reasons People Move
Sometimes people are "pushed" to emigrate because of negative factors in their home country. Perhaps there is a shortage of jobs or a lack of good farmland. At other times, people are "pulled," or attracted, to the new country by such things as job opportunities. **Refugees** are people who are forced to flee their country in order to escape wars, persecution, or natural disasters.

Population Movement (continued)

Impact of Migration

Migration can have positive and negative effects on both the home country and the new country. Emigration might ease overcrowding, but families can be divided. If skilled workers leave a country, the economy can suffer. Immigration can enrich the culture of the new country with different music, art, foods, and traditions. However, some native-born citizens may resent the immigrants. This can lead to unjust treatment and even violence.

Section Wrap-Up

Answer these questions to check your understanding of the entire section.

1. **Explaining** What are two causes of population growth? Explain each.

   _____
   _____

2. **Speculating** Why do geographers study the population density of a country in addition to its total population?

   _____
   _____

Expository Writing

In the space provided, write a paragraph explaining how your community might be affected if a large number of people emigrated from the area or immigrated to it.

_____
_____
_____
_____
_____
_____
_____

Chapter 3, Section 1

Chapter 3, Section 2 (Pages 82–89)
Global Cultures

Big Idea

The world is made up of different cultures with some common traits. As you read, complete the diagram below by identifying the elements of culture.

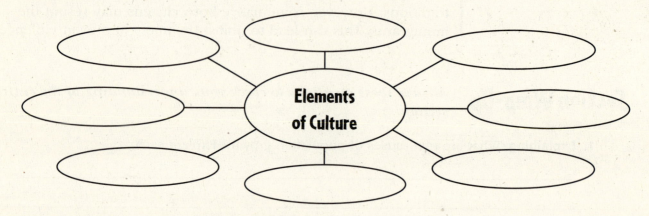

Notes | Read to Learn

What Is Culture? (pages 83–86)

Listing

What social groups are included in a culture?

_____

_____

_____

Culture is the way of life of a group of people who share similar beliefs and customs. Some of the shared cultural elements include language, religion, history, daily life, arts, and government.

Social Groups

Scientists study culture by looking at different groups in a society, such as the young and the old, and males and females. In all cultures, the family is the most important social group. Another type of group is an **ethnic group.** People in an ethnic group share a language, history, religion, and physical traits.

Language, Religion, History, Daily Life, Arts

Sharing a language unifies people in a culture. A language can be spoken in different ways, however. A **dialect** is a local form of a language with unique vocabulary and pronunciation. More than 2,000 languages are spoken throughout the world.

Five major religions and hundreds of other religions are practiced on Earth. Religious beliefs help people answer questions about life's meaning. A shared history also helps a culture define what is important.

What Is Culture? (continued)

Labeling

Indicate whether the types of government below are limited or unlimited.

1. democracy
   _____

2. dictatorship
   _____

3. monarchy
   _____

Cultures vary in the types of food the people eat, the clothing they wear, and the types of homes they build. In addition, music, painting, and other arts reflect what a culture thinks is beautiful and meaningful.

Government and Economy

Governments create rules so people can live together without conflict. In a **democracy,** the government is limited, and the people hold power. A **dictatorship** is a type of unlimited government in which a leader or dictator rules the country, often by force. A government in which a king or queen inherits power is called a **monarchy.** Monarchs at one time had unlimited power. Today most monarchies are constitutional, or share power with elected officials.

A country's economy includes the ways people earn a living, use resources, and trade with other countries. If an economy is successful, most of the people have a good quality of life.

Cultural Change (pages 86–87)

Identifying

Identify four ways that cultures spread.

1. _____
2. _____
3. _____
4. _____

Cultures change over time. These changes can result from technological improvements or from the influence of other cultures.

Inventions and Technology

After 8000 B.C., people learned to farm and began to settle in one place. Historians call this change the Agricultural Revolution. It allowed people to create **civilizations,** or highly developed cultures. In the 1700s, countries began to change from farming societies to industrialized societies. They used machines to make goods.

More recently, computers and other inventions have changed the way people work and communicate. Medical technology has led to people living longer. These advancements have caused cultures to change.

Cultural Diffusion

Influences from other cultures also spark cultural change. Ideas, languages, and customs are spread from one group to another through **cultural diffusion.** People have been exposed to other cultures through trade, migration, and conquest. Today the Internet, television, and movies spread cultural ideas faster than ever.

Chapter 3, Section 2

Notes | Read to Learn

Regional and Global Cultures (pages 88–89)

Finding the Main Idea

What is the main idea of this subsection?

_____
_____
_____
_____

Geographers divide the world into physical regions and into cultural regions. A **culture region** is an area made up of several countries that share cultural traits.

Culture Regions

The countries within a culture region usually have similar languages, histories, and ethnic groups. They also tend to share the same religion and form of government. For example, Canada and the United States make up a culture region.

Global Culture

Increased communication has torn down barriers between culture regions. The result is **globalization,** or a worldwide culture. A key feature of globalization is an interdependent economy. Countries now depend upon one another for resources and markets.

Section Wrap-Up

Answer these questions to check your understanding of the entire section.

1. **Defining** What is a culture region? Give an example of one.
   _____
   _____

2. **Determining Cause and Effect** Why is globalization occurring?
   _____
   _____

Write a paragraph explaining what social groups are in general, and then describe all the social groups to which you belong.

_____
_____
_____
_____
_____

Chapter 3, Section 3 (Pages 92–96)

Resources, Technology, and World Trade

Big Idea

Nations of the world trade resources, creating global interdependence. As you read, complete the chart below. List three examples of each type of natural resource.

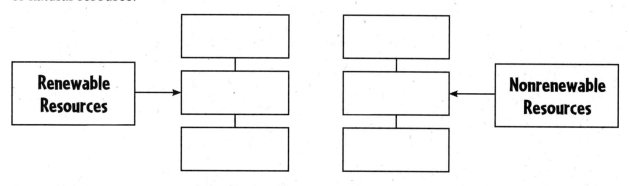

Read to Learn

Nativism Resurges (page 610)

Paraphrasing

Complete these sentences by filling in the blank and then circling "can" or "cannot."

1. _____ resources are available in an unlimited supply and can/cannot be used up.

2. _____ resources are available in a limited supply and can/cannot be used up.

People use Earth's **natural resources** to meet their needs. Some examples of natural resources are soil, trees, wind, and oil. Natural resources provide food, shelter, goods, and energy.

There are two basic types of natural resources. **Renewable resources** cannot be used up, or they can be replaced. The sun, wind, and water cannot be used up. Forests can grow again. Rivers, the wind, and the sun can be used to produce electricity. They are valuable sources of energy.

Nonrenewable resources are limited. When these resources are used up, they cannot be replaced. Minerals such as iron ore and gold are nonrenewable resources. Fossil fuels—oil, coal, and natural gas—are also available in limited amounts, and, in fact, could run out. Fossil fuels are important sources of energy. They are used to heat homes, run cars, and generate electricity.

Chapter 3, Section 3

Economies and Trade (pages 94–96)

Identifying
What are the four kinds of economic systems?

1. _____
2. _____
3. _____
4. _____

Classifying
How do geographers classify countries according to the strength of their economies?

1. _____
2. _____
3. _____

Defining
Explain the meaning of the terms export *and* import *by using each word in a sentence.*

_____
_____
_____
_____
_____
_____

Economic Systems

Every society has to make economic decisions. Societies use an **economic system** to decide what goods and services to produce, how to produce them, and who will receive them.

Four kinds of economic systems exist. In a traditional economy, individuals produce goods based on traditions and customs—just like their parents and grandparents did. In a command economy, the government—not individuals—makes economic decisions. In a market economy, people and businesses decide what to make and buy, and prices are based on supply and demand. The fourth and most common type of economic system is a mixed economy.

Developed and Developing Countries

Geographers define a country by how developed its economy is. A **developed country** has some agriculture, much manufacturing, and service industries like banking and health care. Workers in developed countries generally have high incomes. The economy relies on new technologies.

Developing countries depend mainly on agriculture and have little industry. Workers generally have low incomes.

Newly industrialized countries are in the process of becoming more industrial. They are on the way to becoming developed countries.

World Trade

Resources are not evenly distributed throughout the world. Therefore, trade is important. Nations **export,** or sell to other countries, extra resources or products. Nations also **import,** or buy from other countries, resources they do not have or products they cannot make (or cannot make as cheaply).

Barriers to Trade

Trade impacts a country's economy, so governments take actions to manage international trade. Governments use trade barriers to encourage citizens to buy products made in their own country. Some governments place taxes called **tariffs** on imported goods to make those items more expensive. Another barrier to trade is a **quota,** which is a limit on the number of specific products that can be imported from a particular country.

Free Trade

Many countries are pushing for **free trade,** or the removal of tariffs and quotas. In 1992 Canada, the United States, and Mexico signed the North American Free Trade Agreement (NAFTA). This treaty removed most of the trade barriers among these countries.

Notes | Read to Learn

Economies and Trade (continued)

Interdependence and Technology

Increased trade among countries has led to globalization. The world's people and economies have become linked together. Because of their economic **interdependence,** countries rely on each other for goods, services, and markets in which to sell their products. When economies are linked, something that happens in one location can have a global effect. For example, a drought in one country might reduce the amount of crops it can grow and sell to others, leading to shortages in countries that want to buy those crops.

Section Wrap-Up

Answer these questions to check your understanding of the entire section.

1. **Applying** Geographers define a country by how developed its economy is. Which label would be the most appropriate for the United States? Why?

   _____
   _____

2. **Drawing Conclusions** What are two trade barriers, and why might a government want to use them?

   _____
   _____
   _____

Informative Writing

In the space provided, write a short story that explains how a war in one part of the world impacts the economy of another country far away.

_____
_____
_____
_____

Chapter 3, Section 3

Chapter 4, Section 1 (Pages 116–122)
Physical Features

Big Idea

Geographers organize the Earth into regions that share common characteristics. In the Venn diagram below, compare landforms in the eastern, western, and interior parts of the United States and Canada.

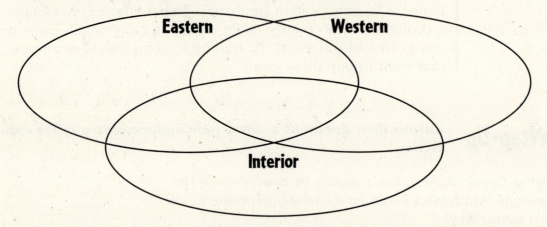

 **Notes** | **Read to Learn**

Major Landforms (pages 117–119)

Defining

Define the terms **contiguous** *and* **megalopolis** *by using each word in a sentence.*

_____
_____
_____
_____
_____

The United States and Canada cover most of North America. The region is bordered by the Arctic Ocean in the north, the Atlantic Ocean in the east, the Gulf of Mexico in the southeast, and the Pacific Ocean in the west. Canada, the second-largest country in the world, stretches across most of the northern part of North America. The United States, the third-largest country, covers the middle part. The 48 states here are **contiguous,** or joined together inside a common boundary. Alaska and Hawaii are separate. Alaska is connected to Canada, and Hawaii lies in the Pacific Ocean about 2,400 miles from California.

A variety of landforms cover the region. Broad lowlands are found along the Atlantic and Gulf coasts. Rocky soil in the Northeast limits farming, but excellent harbors led to large shipping ports there. In the United States, cities and suburbs along the Atlantic form a **megalopolis,** or connected area of urban communities. Moving west, highland areas rise, including the Appalachian Mountains. The Appalachians run from eastern Canada to Alabama, and their rounded peaks have eroded over time. Still, rich coal deposits are mined here.

28   Chapter 4, Section 1

Major Landforms (continued)

Locating
What physical features are located among the cordillera?

_____

_____

_____

Interior lowlands are found west of the Appalachians. The horseshoe-shaped Canadian Shield wraps around Hudson Bay. Rich minerals are mined in this cold, rocky region. South of the Shield spread the Central Lowlands with grassy hills, thick forests, fertile farmland, and the Great Lakes. The Mississippi River and large cities—Chicago, Detroit, Cleveland, and Toronto, for example—are located in the Central Lowlands. West of the Mississippi stretches the Great Plains—a vast **prairie** of rolling grasslands with rich soil.

A **cordillera,** or group of mountain ranges that run side by side, towers over much of the west. The Rocky Mountains begin in Alaska and run south to New Mexico. Mountains near the Pacific coast include the Sierra Nevada, the Cascades, the Coast Ranges, and the Alaska Range. Mount McKinley—North America's highest point—rises 20,320 feet in the Alaska Range. Among the ranges lie dry basins, high plateaus, and magnificent **canyons,** or deep valleys with steep sides. The most famous is the Grand Canyon of the Colorado River.

Bodies of Water (pages 119–120)

Specifying
Where do rivers on each side of the Continental Divide flow?

West:

_____

_____

East:

_____

_____

_____

Numerous freshwater rivers and lakes are found in this region. Many are **navigable,** or wide and deep enough to allow ships to travel on them. The Great Lakes—Superior, Huron, Erie, Michigan, and Ontario—are the world's largest group of freshwater lakes. **Glaciers,** or giant ice sheets, carved them thousands of years ago. The connected lakes flow into Canada's St. Lawrence River, which empties into the Atlantic Ocean. Cities on these lakes—as well as Quebec, Montreal, and Ottawa—depend on the St. Lawrence Seaway for shipping.

The longest river in North America is the Mississippi River. It begins in Minnesota and flows 2,350 miles before emptying into the Gulf of Mexico. Products from inland port cities, such as St. Louis and Memphis, are shipped down the Mississippi to foreign ports.

Many rivers, including the Colorado and Rio Grande, flow from the Rocky Mountains. The high ridge of the Rockies forms the Continental Divide. A **divide** is a point that determines which direction rivers will flow. Rivers to the east of the Continental Divide flow toward the Arctic or Atlantic Oceans or into the Mississippi. Rivers to the west of the divide flow toward the Pacific Ocean.

Chapter 4, Section 1

Notes | Read to Learn

Natural Resources (pages 121–122)

Finding the Main Idea

As you read, write down the main idea of this subsection.

_____
_____
_____
_____

Abundant resources have enabled the United States and Canada to build strong industrial economies. Energy resources include oil and natural gas. Texas and Alaska have large oil reserves. The United States must import oil to meet all of its needs, however. In Canada, oil and gas reserves are found near the province of Alberta. Both countries also have rich coal deposits, and hydroelectric power is generated from rivers. Niagara Falls is a major source of hydroelectricity.

The region's mineral resources are plentiful. The Rocky Mountains yield iron ore, gold, silver, and copper. Iron ore, nickel, gold, copper, and uranium are mined in the Canadian Shield.

Rich soil provides excellent farmland in parts of the region. The types of crops grown are determined by local climates, but irrigation allows even dry areas to grow many products. In California alone, farmers grow more than 200 different crops.

Forests supply timber for lumber and wood products. Large fishing industries thrive in coastal waters. The Grand Banks, once a rich fishing ground, has suffered from overfishing, however.

Section Wrap-Up

Answer these questions to check your understanding of the entire section.

1. **Explaining** In what way is the Mississippi River important to trade for the United States?

   _____
   _____

2. **Categorizing** The natural resources of the region can be divided into three broad categories. List the region's resources in their proper categories.

Energy	Mineral	Other Resources

On a separate sheet of paper, write a paragraph describing a mountain range, a waterfall, or a canyon.

Chapter 4, Section 2 (Pages 124–128)
Climate Regions

Big Idea

The physical environment affects how people live. As you read, complete the chart below by organizing key facts about three different climate zones in the region.

Climate Zone	Location	Description
1.		
2.		
3.		

Read to Learn

A Varied Region (pages 125–127)

Explaining

What factors contribute to the desert climate in the inland West?

_____
_____
_____
_____

The region of the United States and Canada extends from the frozen Arctic in the far north to the steamy Tropics in the south. This results in a great variety of climates and vegetation. Most Americans and Canadians tend to live in the moderate climate zones found in the middle latitudes.

Tundra and subarctic climates are found in the northern parts of Alaska and Canada. The winters are long and cold, and the summers are short and cool. The tundra climate along the Arctic Ocean prevents the growth of trees and most plants. However, dense forests of evergreen trees grow farther south in the subarctic region.

Moist ocean winds affect the climate along the Pacific coast. A marine west coast climate is found from southern Alaska to northern California. This climate zone has mild temperatures and abundant rainfall. Southern California has a Mediterranean climate with warm, dry summers and mild, wet winters.

The inland West has a desert climate with hot summers and mild winters. Humid ocean winds are blocked from reaching this area by the mountains. In addition, hot, dry air gets trapped between the Pacific ranges and the Rocky Mountains.

A Varied Region (continued)

Applying

List the climates that support each type of plant life below.

1. *Prairie grasses and grains:*
_____

2. *Variety of forests:*
_____

3. *Wetlands and swamps:*
_____

4. *Rain forests:*
_____

The eastern side of the Rockies has a steppe climate. Long periods without rain, or **droughts,** provide challenges, especially to farmers and ranchers in this area. A growing population here also strains water resources.

Much of the Great Plains has a humid continental climate. Moist winds blow from the Gulf of Mexico and from the Arctic. Winters are usually cold and snowy, and summers are hot and humid. Enough rain falls to allow prairie grasses and grains to grow. The Great Plains may experience drought, however. In the 1930s, dry weather and winds turned the Great Plains into a wasteland called the Dust Bowl. The soil has since been restored through better farming methods.

The northeastern United States and parts of eastern Canada have a humid continental climate. In contrast, a humid subtropical climate is found in the southeastern United States, along with wetlands and swamps in some areas. These two climates have similar temperatures in the summer, and both have a variety of forests. In the winter, however, icy Arctic air makes the Northeast much colder.

Two parts of the United States have tropical climates. Southern Florida is in a tropical savanna zone, with hot summers and warm winters. Rain falls mostly during the summer. Hawaii is also tropical, with warm temperatures and enough rainfall to support tropical rain forests.

Natural Hazards (pages 127–128)

Identifying

Identify three weather-related natural hazards.

_____

_____

Severe storms and other types of natural hazards pose challenges for the United States and Canada. One type of severe weather is a **tornado,** or a windstorm in the form of a funnel-shaped cloud. If the tornado touches the ground, its strong winds can knock down buildings and trees as well as lift up and move large objects. The central United States has been nicknamed "Tornado Alley" because more tornadoes form here each year than any other place in the world.

Hurricanes are wind systems that form over the ocean in tropical areas. They bring violent storms with heavy rains. Not only are the winds damaging, but hurricanes also can cause storm surges, when high levels of seawater flood low coastal areas. Hurricane season usually lasts from June to September. In August 2005, Hurricane Katrina struck the coast along the Gulf of Mexico. One of the most damaging hurricanes in history,

 **Read to Learn**

Natural Hazards (continued)

Determining Cause and Effect

What natural hazards are most likely to occur along the Pacific Coast? Why?

_____
_____
_____
_____

Katrina killed more than 1,800 people, wiped out hundreds of thousands of homes, and destroyed entire towns.

Severe winter storms called **blizzards** can last several hours and combine high winds with heavy snow. The snow can fall so heavily that people cannot see far, causing "white-out" conditions and making it dangerous to be outside. The snow and winds can knock down power lines and trees.

Not all natural hazards in the region are caused by the weather. Earthquakes can occur anywhere in the region, but they generally occur along the Pacific coast, where fault lines of tectonic plates meet. Volcanoes also are found where tectonic plates meet—in the Pacific ranges, southern Alaska, and Hawaii. Most of the volcanoes in this region are dormant, but several of Hawaii's volcanoes are active.

Section Wrap-Up *Answer these questions to check your understanding of the entire section.*

1. **Analyzing** Where do most Americans and Canadians live? Why?

   _____
   _____

2. **Describing** What are the characteristics of a marine west coast climate and a Mediterranean climate? Where are these two climates located in this region?

   _____
   _____

 *On a separate sheet of paper, write a short story about a tornado, hurricane, blizzard, earthquake, or volcanic eruption as if you were experiencing the natural hazard.*

Chapter 4, Section 2

Chapter 5, Section 1 (Pages 134–141)
History and Governments

Big Idea

The characteristics and movement of people impact physical and human systems. As you read, fill in the time line below with key events and dates in the history of the United States and Canada.

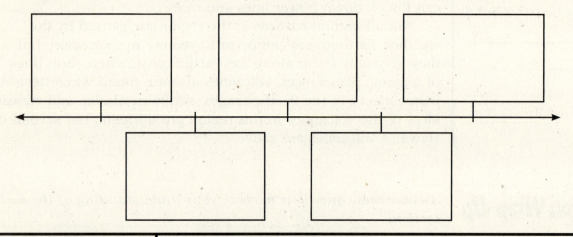

Read to Learn

History of the United States (pages 135–138)

Naming

Name the three countries that established colonies in the Americas.

1. _____
2. _____
3. _____

The first people to settle the Americas were hunters from Asia who traveled with herds across a land bridge between Siberia and Alaska about 15,000 years ago. Descendants of these first settlers are called Native Americans.

Christopher Columbus reached the Americas in 1492. Soon Spain, France, and Great Britain established American **colonies,** or overseas settlements with ties to a parent country. In 1763 Great Britain gained control of France's colonies. The people in Britain's 13 coastal colonies grew resentful of British taxes and trade policies. They declared independence in 1776 and fought against British troops. Britain recognized American independence in 1783, and the United States was established.

During the 1800s, the country expanded to the Pacific Ocean, often by **annexing,** or declaring ownership of, areas of land. This expansion brought great suffering to Native Americans.

The population and economy also expanded during the 1800s. Millions of Europeans immigrated to the United States. New machines helped farmers, and the factory system produced many goods. Roads, canals, steamboats, and railroads moved goods to market quickly.

History of the United States (continued)

Identifying Central Issues

Over what major issue was the Civil War fought?

The country became divided, however. Southern states relied on agriculture and the work of enslaved Africans. People in the north were more industrialized and criticized slavery. In 1861 several southern states withdrew from the United States. The Civil War was fought to reunite the country. Slavery ended, but racial tensions continued.

The United States became a world leader during the 1900s, fighting in World War I and World War II. After World War II, the United States and the Soviet Union struggled for world leadership in the Cold War. During this period, struggles occurred at home. Native Americans, African Americans, Latino Americans, and women sought equal rights. Since 2000, **terrorism,** or violence against civilians to reach political goals, has become a new threat. On September 11, 2001, Muslim terrorists seized four passenger planes and crashed them into New York City and Washington, D.C. About 3,000 people died.

History of Canada (pages 138–139)

Locating

Where was New France located in Canada?

Stating

What caused the Canadian colonies to unite?

Canada also was first settled by Native Americans. Viking explorers from Scandinavia lived briefly on the Newfoundland coast about A.D. 1000, but they eventually left.

France and England both claimed areas of Canada in the 1500s and 1600s. The French ruled New France—the area around the St. Lawrence River and the Great Lakes. They also established the cities of Quebec and Montreal. The French became wealthy by trading with Native Americans for beaver furs, which they sold in Europe. England and France continued to fight for territory around the world in the 1600s and 1700s. By the 1760s, Britain gained control of New France.

Fearing that the United States would try to take them over, most of the colonies in Canada joined together in 1867 to become the Dominion of Canada. As a **dominion,** Canada had its own government to take care of local matters, but Britain controlled Canada's relations with other countries. The colonies became provinces, which are similar to states. Today Canada has ten provinces and three territories. The culture and language of the province of Quebec is French.

During the 1900s, Canadians fought alongside Americans and the British in the two World Wars. Canada became fully independent in 1982. The country faces the possibility that Quebec will separate and become independent.

Chapter 5, Section 1

Notes | Read to Learn

Governments of the United States and Canada (pages 140–141)

Summarizing

Complete these sentences.

In a _____ _____, voters elect their leaders.

A _____ _____ is a type of democracy in which elected representatives choose a _____.

The United States and Canada each have a **representative democracy**—voters elect leaders who make and enforce the laws. The two systems are different in several ways, however.

The U.S. Constitution, written in the 1780s, provides the basic plan for how our national government is set up and works. Power is divided among three branches: executive, legislative, and judicial. The Constitution also set up a system of **federalism,** in which power is divided between the national and state governments.

Over the years, **amendments,** or additions, were added to the U.S. Constitution. The first 10 amendments, added in 1791, are known as the Bill of Rights. They guarantee basic freedoms.

Canada is a **parliamentary democracy.** People elect representatives to a lawmaking body called Parliament. The members of Parliament then choose a prime minister to rule.

Canada also has a federal system, with responsibilities divided between the national and provincial or territorial governments. Canada's Charter of Rights and Freedoms is similar to the U.S. Bill of Rights. It protects the liberties of Canadian citizens.

Section Wrap-Up

Answer these questions to check your understanding of the entire section.

1. **Explaining** What struggle occurred in the United States during the Cold War period? Who was involved in this struggle?

   _____

2. **Comparing and Contrasting** Compare and contrast the governments of the United States and Canada in the Venn diagram below.

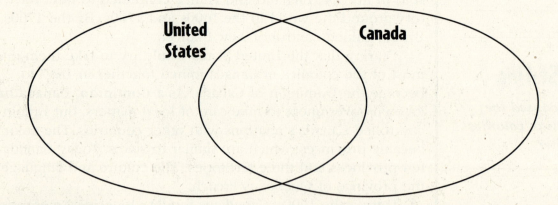

The Bill of Rights guarantees certain freedoms to the people. On a separate sheet of paper, write a paragraph explaining the freedoms that you value most.

Chapter 5, Section 1

Chapter 5, Section 2 (Pages 144–150)
Cultures and Lifestyles

Big Idea

Culture influences people's perceptions about places and regions. As you read, list key facts in the chart below about the cultures and lifestyles of the United States and Canada.

	United States	Canada
Language		
Art and Literature		
Leisure Time		

 **Notes** **Read to Learn**

Cultures and Lifestyles of the United States (pages 145–147)

Displaying

Create a circle graph to reflect the ethnic breakdown of the U.S. population.

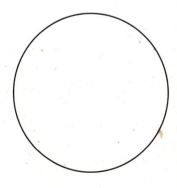

The United States has about 300 million people from different ethnic backgrounds. Early immigrants came mainly from Great Britain and Ireland. In the late 1800s, immigrants arrived from other areas of Europe, China, Japan, Mexico, and Canada. So many diverse backgrounds led some Americans to worry about cultural change. In 1882 and 1924, laws were passed that **banned,** or blocked, most immigration from China and many other countries. Immigration slowed. However, changes in U.S. laws in the 1960s increased immigration again. By 2000, nearly half of the immigrants came from Latin America and Canada, and one-third came from Asia.

People of European origin still make up two-thirds of the population. Latinos, or Hispanics, make up 15 percent and are the fasting-growing ethnic group. African Americans comprise 12 percent; Asian Americans, 4 percent; and Native Americans, 1 percent.

English is the primary language, followed by Spanish. Chinese, French, Vietnamese, Tagalog, German, and Italian are each spoken by more than 1 million people.

Most Americans follow a form of Christianity. Other religions practiced include Judaism, Islam, Buddhism, and Hinduism.

Chapter 5, Section 2   37

Cultures and Lifestyles of the United States (continued)

Drawing Conclusions

Why do you think Americans move to the Sunbelt?

_____
_____
_____

Artists, writers, and musicians developed distinctly American styles. Native Americans used materials from the environment to create their works, including wooden masks and pottery. Later artists focused on the beauty of the landscape or the gritty side of city life. Common themes in literature are the diversity of the people and the history and landscapes of different regions. American musicians have created many styles—folk, country, blues, jazz, rock and roll, rap, and hip-hop.

Most Americans live in cities or **suburbs,** or smaller communities around a larger city. Since the 1970s, the fastest-growing regions have been the South and Southwest—called the Sunbelt. Americans lead the world in the ownership of homes, cars, and personal computers. They also have the highest Internet usage rate. Leisure activities include watching movies and television and playing sports. Important holidays include Thanksgiving, the Fourth of July, and celebrations based on religion.

Cultures and Lifestyles of Canada (pages 149–150)

Identifying

What percentage of Canadians are of French ancestry?

_____

Summarizing

Why do many people in Quebec want to be separate from Canada?

_____
_____
_____
_____

Canada, like the United States, is made up of immigrants with many different cultures. In Canada, however, physical distances and separate cultures have led many people to feel more attached to their region than to the nation as a whole.

About 25 percent of Canadians are of French descent, and 25 percent have British origins. Another 15 percent are of other European backgrounds. Canada has more than a million **indigenous** people, or descendants of the area's first inhabitants. They are referred to as the "First Nations."

Canada is a **bilingual** country with two official languages—English and French. Many French speakers in Quebec want to become independent to better preserve their language and culture. Another cultural group that desired self-rule was the Inuit, a northern indigenous people. In 1999 Canada created the territory of Nunavut for them. There, the Inuit mostly govern themselves.

Early indigenous artists carved figures from stone and wood, made pottery, or were weavers. Canadian artists today are influenced by European and indigenous cultures. Music has changed over time, from the religious rituals of the indigenous people, to Irish and Scottish ballads in the 1700s, to pop and rock today. Movies and theater also are popular in Canada.

Foods vary by region. Seafood is common in the Atlantic Provinces. Quebec offers French cuisine. Ontario features Italian

Cultures and Lifestyles of Canada (continued)

Specifying

What territory was created for the Inuit?

_____

and Eastern European foods, which reflect the immigrants who settled there. Along the Pacific, British Columbia is known for salmon and Asian foods.

Canadians enjoy hockey—a sport that began in Canada. They also play lacrosse—originally a Native American game. Outdoor activities such as hunting and fishing are enjoyed as well. Canada's independence is celebrated on July 1. Like Americans, the people of Canada celebrate Thanksgiving in the fall.

Section Wrap-Up

Answer these questions to check your understanding of the entire section.

1. **Naming** What religion do most Americans practice? What other religions are practiced in America?

   _____
   _____

2. **Defining** What does it mean to say that Canada is bilingual?

   _____
   _____

In the space provided, write a paragraph explaining how immigration has led to diversity in America.

_____
_____
_____
_____
_____
_____
_____

Chapter 5, Section 2

Chapter 6, Section 1 (Pages 158–162)
Living in the United States and Canada Today

Big Idea

Places reflect the relationship between humans and the physical environment. As you read, complete the diagram below. List the U.S. economic regions and provide key facts about each.

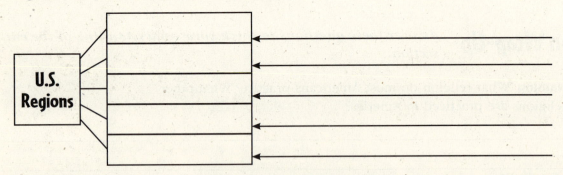

 **Notes** **Read to Learn**

Economic Regions (pages 159–161)

Defining

Define the words profit and stock by using each one in a sentence.

_____
_____
_____
_____
_____
_____
_____

The United States and Canada have **free market** economies. People are able to buy, sell, and produce whatever they want with limited government involvement. They also can work whenever they want. Business owners produce the items they believe will have the highest **profits**, or make the most money after expenses are paid. Consumers look for the best products at the lowest prices.

Individuals can invest in businesses by buying **stock**, which represents part ownership in a company. Owning stock allows investors to share a company's profits. If the company fails, however, the stock becomes worthless. Individuals may choose to save their money in a bank. This is safer than buying stock but has less of a chance for a high financial payoff.

The United States is divided into five economic regions. Each region specializes in making products from its available resources. The Northeast region is made up of large urban areas. With little fertile soil, this region focuses on business. New York City—the country's most populous city—has many financial and media companies. Boston, Massachusetts, is a center of **biotechnology** research, or the study of cells to improve health.

40   Chapter 6, Section 1

Economic Regions (continued)

Listing

List six industries found in the Interior West.

1. _____
2. _____
3. _____
4. _____
5. _____
6. _____

The rich soil of the Midwest region allows farmers to grow crops such as corn, wheat, and soybeans. The region also has deposits of iron ore, coal, lead, and zinc. Cities such as Detroit and Cleveland once were centers of auto and steel manufacturing. Factories became outdated, however, and many closed. Thousands of jobs were lost.

The South has rich soil and much agriculture. Recently, the South also has experienced growing cities and industries. Textiles, electrical equipment, computers, and airplane parts are manufactured in Houston, Dallas, and Atlanta. Texas, Louisiana, and Alabama produce oil. Tourism and trade thrive in Florida.

The Interior West has long been supported by mining, ranching, and lumbering. Information technology and service industries have grown rapidly in Denver and Salt Lake City. Beautiful scenery attracts tourists to places in this region, such as Phoenix and Albuquerque.

Fruits and vegetables grow in the fertile Pacific region. Hawaii has sugarcane, pineapples, and coffee. Resources such as fish, timber, mineral deposits, and oil reserves also are plentiful. Workers in California and Washington build planes and develop software. Los Angeles is world famous for its movie industry.

Regions of Canada (pages 161–162)

Identifying

Identify the economic regions of Canada.

1. _____
2. _____
3. _____
4. _____

Canada also has distinct economic regions and a free market economy. In Canada, however, the government has a more direct role in providing services. It provides health care for citizens and regulates broadcasting, transportation, and power companies.

The Atlantic Provinces—Nova Scotia, Prince Edward Island, New Brunswick, and Newfoundland and Labrador—once had a profitable fishing industry. Overfishing, however, caused the industry to decline. As a result, many workers moved into manufacturing, mining, and tourism. Halifax, Nova Scotia, is a major shipping center.

The Central and Eastern region is made up of the provinces of Quebec and Ontario. The paper industry and hydroelectric power are two important industries in Quebec. Montreal is a major port on the St. Lawrence River, as well as a leading financial and industrial center. Foreign businesses are reluctant to invest in Quebec's economy, though, because it wants to separate from Canada.

Ontario is the wealthiest province and has the largest population. Its economic activities include agriculture, manufacturing,

Chapter 6, Section 1

 **Read to Learn**

Regions of Canada (continued)

Summarizing

Summarize Ontario's importance by providing five important details about it below.

1. _____
2. _____
3. _____
4. _____
5. _____

forestry, and mining. The city of Toronto is a major business and finance center. Immigrants from 170 countries have made Ontario their home.

Three provinces in the West—Manitoba, Saskatchewan, and Alberta—are known for farming and ranching. Wheat is a major export. This region also has some of the world's largest reserves of oil and natural gas. Extensive forests cover British Columbia. Lumber is used to produce **newsprint**, the type of paper used for printing newspapers. British Columbia's economy also includes mining, fishing, and tourism. Vancouver is Canada's main Pacific port.

Canada's northern region—the Yukon Territory, the Northwest Territories, and Nunavut—takes up one-third of the country. Only about 25,000 people live there, however. Many are indigenous peoples. The North has mineral deposits of gold and diamonds.

Section Wrap-Up

Answer these questions to check your understanding of the entire section.

1. **Explaining** What is a free market economy, and how does Canada's differ from that of the United States?

   _____
   _____

2. **Analyzing** In what way is the major economic activity in Canada's Atlantic Provinces changing? Why?

   _____
   _____

 *Imagine that you are a factory worker in the Midwest region of the United States. The factory's owners are considering shutting down the plant. On a separate sheet of paper, write a paragraph explaining why the factory should remain open.*

Chapter 6, Section 2 (Pages 168–172)
Issues and Challenges

Big Idea

Cooperation and conflict among people have an effect on the Earth's surfaces. As you read, complete the outline below. Write each main heading on a line with a Roman numeral, and list important facts below it.

I. First Main Heading _____
   A. Key Fact 1 _____
   B. Key Fact 2 _____
II. Second Main Heading _____
   A. Key Fact 1 _____
   B. Key Fact 2 _____

Notes | Read to Learn

The Region and the World (pages 169–171)

Discussing

Write a brief explanation of free trade, and identify one action that the United States and Canada have taken to support it.

_____
_____
_____
_____
_____

The United States and Canada have large, productive economies. They trade with countries throughout the world. The United States, in fact, has the world's largest economy and is a leader in world trade.

The United States and Canada support free trade. They want to remove barriers in order to ease trade between countries. In 1994 the United States, Canada, and Mexico signed the North American Free Trade Agreement (NAFTA), eliminating most trade restrictions among the three countries. Today Canada is the largest trading partner of the United States, and Mexico is the second largest.

Major U.S. exports include chemicals, farm products, and manufactured goods, as well as raw materials like metals and cotton. Canada has many of the same exports, as well as seafood and timber.

Both countries also import many goods. The United States imports most of its energy resources, particularly oil. Suppliers of oil include Canada, Mexico, Venezuela, Saudi Arabia, Nigeria, and Angola.

Notes | Read to Learn

The Region and the World (continued)

Specifying

Write three ways the United States and Canada participate in the UN.

1. _____
2. _____
3. _____

American consumers buy many foreign products. The United States has a huge **trade deficit**—it spends hundreds of billions of dollars more on imports than it earns from exports. This has occurred because some countries set the prices of their products low. Low prices encourage sales in the United States. At the same time, some countries put **tariffs,** or taxes, on imports to protect their own industries. Tariffs make U.S. products more expensive and reduce their sales abroad, which hurts American companies and their workers.

In contrast, Canada has a **trade surplus,** meaning that it earns more from exports than it spends on imports. Canada has a smaller population than the United States, so its energy needs are less costly.

Since the early 2000s, the United States and Canada have joined other countries to combat terrorism and violence. On September 11, 2001, terrorists attacked sites in the United States. Since then, the United States and Canada have increased border security and joined international efforts to prevent terrorist attacks.

Both the United States and Canada have important roles in the United Nations (UN), the world organization that promotes cooperation among countries in settling disputes. They provide funding, participate in agencies that provide international aid, and send soldiers to serve in UN forces.

Environmental Issues (pages 171–172)

Displaying

Complete the diagram below.

Causes
1.
2.

↓

Lower Water Levels

Effects
1.
2.
3.

The United States and Canada share environmental concerns. As people burn coal, oil, and natural gas, chemicals are released that pollute the air. Air pollution mixes with water vapor to make **acid rain,** or rain that has high amounts of chemical pollutants. Acid rain can harm trees, waterways, and the stone used in buildings. The two countries are taking steps to limit the amount of chemicals released into the air.

Another environmental concern is global warming. Some scientists think that warmer temperatures will change weather patterns, which may cause drought or melt polar ice caps. Low-lying areas like Florida may be flooded. To address this concern, Canada has passed laws to limit the amount of fossil fuels that can be burned. The United States is researching new, less harmful energy sources.

The water levels of the Great Lakes have dropped sharply because of climate changes and an increased demand for water. The lower lake levels harm fish and affect the shipping and

Environmental Issues (continued)

Listing

List four effects of urban sprawl.

_____
_____
_____
_____

tourism industries. The governments of both countries have urged conservation.

Another environmental challenge is **brownfields.** These abandoned places, such as factories and gas stations, contain dangerous chemicals. Until these chemicals are cleaned up, new development cannot occur at the sites. The governments of both countries have given money to communities to help clean up their brownfields.

The spread of human settlement into natural areas is called **urban sprawl.** Urban sprawl has led to the loss of farmland and wilderness, as well as increased traffic jams, pollution, and strains on water and other resources.

Section Wrap-Up

Answer these questions to check your understanding of the entire section.

1. **Determining Cause and Effect** What actions taken by other countries have resulted in a huge U.S. trade deficit?

   _____
   _____

2. **Describing** How are the United States and Canada addressing global warming?

   _____
   _____

Review the information about the causes of acid rain and develop a course of action that might reduce it. On a separate sheet of paper, write a letter to a member of Congress explaining your idea and requesting that it be put into action.

Chapter 6, Section 2

Chapter 7, Section 1 (Pages 192–196)
Physical Features

Big Idea

Geographic factors influence where people settle. As you read, complete the diagram below. Identify six key landforms in this region.

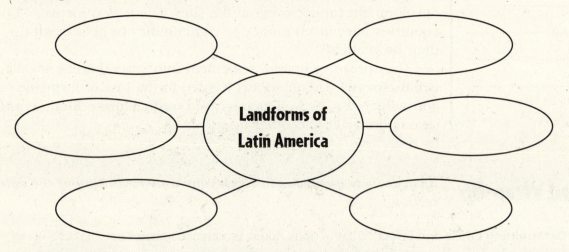

Read to Learn

Landforms (pages 193–194)

Differentiating

List the three groups of islands in the Caribbean and a key fact about each.

1. _____
   _____
   _____

2. _____
   _____
   _____

3. _____
   _____

Latin America is divided into three **subregions,** or smaller areas—Middle America, the Caribbean, and South America. Mexico and Central America make up Middle America. Central America is an **isthmus,** or narrow piece of land that links two larger areas of land. It links North and South America. Middle America has active volcanoes and frequent earthquakes because four tectonic plates meet there. Mountain ranges rise along Mexico's eastern and western coasts, and a high plateau lies between them. Forested mountains also form a backbone through Central America. Narrow, marshy lowlands are found along the Pacific and Caribbean coastlines.

The Caribbean islands, also called the West Indies, are divided into three groups. The Greater Antilles includes the largest islands—Cuba, Hispaniola, Puerto Rico, and Jamaica. The Lesser Antilles is an **archipelago,** or chain of islands, curving from the Virgin Islands to Trinidad. The third group, the Bahamas, is another archipelago. Some Caribbean islands are low-lying. Others, formed by volcanoes, have rugged mountains and fertile volcanic soil.

Landforms (continued)

The 5,500-mile-long Andes ranges and the huge Amazon Basin are major landforms in South America. The Brazilian Highlands that border the Amazon Basin end in an **escarpment,** or series of steep cliffs, that drop to the Atlantic coast. Tropical grasslands called the **Llanos** cover eastern Colombia and Venezuela. Another fertile plain called the **Pampas** stretches through much of Argentina and Uruguay.

Waterways (pages 194–195)

Explaining

Why is the Amazon River important to Latin America?

_____
_____
_____
_____

The longest river in Latin America is the Amazon. It begins in the Andes and flows about 4,000 miles to the Atlantic Ocean. Many **tributaries,** or smaller rivers, flow into the Amazon. The Amazon is used for shipping, and people also rely on it for fish.

The Paraná, Paraguay, and Uruguay Rivers form the second-largest river system in Latin America. These three rivers wind through inland areas and then flow into an **estuary,** the place where river currents meet ocean tides. This estuary—called the Río de la Plata, or "River of Silver"—flows into the Atlantic Ocean. The Orinoco River flows north through Venezuela into the Caribbean Sea.

The largest lake in South America is Venezuela's Lake Maracaibo. Lake Titicaca is located 12,500 feet above sea level high in the Andes between Bolivia and Peru. It is the world's highest navigable lake. Another key waterway in the region is the Panama Canal. This human-made waterway across the Isthmus of Panama provides a shorter route for ships traveling between the Atlantic and Pacific Oceans.

A Wealth of Natural Resources (pages 195–196)

Discussing

What is gasohol, and why does Brazil produce it?

_____
_____

Brazil is the largest country in Latin America and has the most natural resources. Rain forests cover more than half of the country and provide timber, rubber, palm oil, and Brazil nuts. Mineral deposits in Brazil include bauxite, gold, tin, iron ore, and manganese. Brazil does not have large oil reserves. To limit dependence on imported oil, Brazil produces a fuel for cars called **gasohol,** an alcohol made from sugarcane and gasoline.

Venezuela and Mexico produce enough oil and natural gas to meet their needs and to export to other countries. Bolivia and Ecuador also have oil and natural gas deposits. Other minerals

Chapter 7, Section 1

Notes | Read to Learn

A Wealth of Natural Resources (continued)

Identifying

Underline the reasons Nicaragua and Guatemala have difficulty mining gold.

in the region include silver in Mexico and Peru. Venezuela has rich iron ore deposits. Colombia mines emeralds, and Chile exports copper.

The Caribbean islands have few mineral resources. Jamaica is an exception. It has large deposits of bauxite, which is used to make aluminum. Cuba mines nickel, and the Dominican Republic mines gold and silver. Nicaragua and Guatemala in Central America have rich gold deposits. However, political conflicts and transportation difficulties make mining their gold difficult.

Section Wrap-Up

Answer these questions to check your understanding of the entire section.

1. **Explaining** What is the significance of the Panama Canal?

   _____

   _____

2. **Organizing** Complete this chart by listing the country(ies) in which the mineral resources are found.

Minerals	Country(ies)
Bauxite	
Copper	
Emeralds	
Gold	
Iron ore	
Manganese	
Nickel	
Silver	
Tin	

Descriptive Writing

On a separate sheet of paper, write a paragraph describing some of the challenges of living in an area with active volcanoes and frequent earthquakes.

Chapter 7, Section 2 (Pages 198–202)
Climate Regions

Big Idea

The physical environment affects how people live. As you read, use the Venn diagram below to compare and contrast the tropical rain forest and the tropical savanna climate zones.

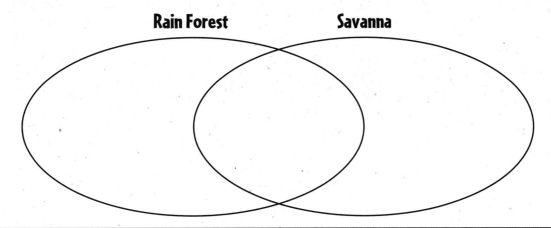

 **Notes** | **Read to Learn**

Hot to Mild Climates (pages 199–201)

Paraphrasing

As you read, complete these sentences.

The Tropics are generally warm year-round because the area receives _____.

Two factors that affect the climate in Latin America are _____ *and* _____.

Most of Latin America is located in the area between the Tropic of Cancer and the Tropic of Capricorn. This area is called the **Tropics.** It receives direct sunlight for much of the year, and the temperatures are generally warm. However, mountain ranges and wind patterns contribute to a variety of climates in the region.

Some Caribbean islands and much of Central America and South America have a tropical wet climate. Temperatures are hot and rainfall is heavy throughout the year. Much of this climate zone is covered by **rain forests,** or dense stands of trees and other plants that thrive on high amounts of rain.

The world's largest rain forest is in South America's Amazon Basin. The trees there grow so close together that their tops form a **canopy,** or an umbrella-like covering of leaves. The canopy blocks most sunlight from reaching the forest floor.

Most Caribbean islands, parts of Middle America, and north central South America have a tropical dry, or savanna, climate zone. Temperatures are hot, and rainfall is abundant, but this climate zone also has a long dry season.

Hurricanes often strike the Caribbean islands from June to November. These storms can cause much damage.

Chapter 7, Section 2   **49**

Notes | Read to Learn

Hot to Mild Climates (continued)

Identifying

Identify two effects of El Niño.

1. _____
_____
_____

2. _____
_____
_____

The areas south of the Tropic of Capricorn have temperate climates. A humid subtropical climate—with short, mild winters and long, hot, humid summers—covers southern Brazil and the Pampas of Argentina and Uruguay. Central Chile has a Mediterranean climate, with dry summers and rainy winters. Farther south is a marine coastal climate, with heavier rainfall throughout the year.

Dry climates are found in some parts of Latin America, such as northern Mexico, coastal Peru and Chile, northeastern Brazil, and southeastern Argentina. Grasslands thrive in the steppe climate, and cacti and shrubs grow in desert zones. One of the driest places on Earth is the Atacama Desert along the Pacific coast of northern Chile. The Andes block rain from reaching this desert. In addition, a cold current in the Pacific Ocean brings only dry air to the coast.

Weather in South America is subject to the El Niño effect. When El Niño occurs, Pacific winds blowing toward land carry heavy rains that lead to flooding along Peru's coast. El Niño also can cause drought in northeastern Brazil.

Elevation and Climate (pages 201–202)

Labeling

Label the four altitude zones in the Andes.

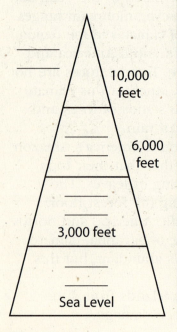

A place's height above sea level, called **altitude**, affects its climate. Higher altitudes have cooler temperatures, even in warm regions such as the Tropics. In South America, the Andes have four altitude zones of climate.

The *tierra caliente,* or "hot land," is named for the hot and humid areas that are near sea level. Farmers here grow bananas, sugarcane, and rice. The *tierra templada,* or "temperate land," is found from 3,000 feet to 6,000 feet. It is moist and pleasant, with mild temperatures. *Tierra templada* is the most densely populated climate zone. People grow corn, wheat, and coffee in this zone.

The *tierra fría,* or "cold land," extends from 6,000 feet to 10,000 feet. Average temperatures in this zone can be as low as 55°F. Crops that do well in these conditions include potatoes, barley, and wheat. The highest altitude, starting at about 10,000 feet, is called *tierra helada,* or "frozen land." Conditions are harsh, vegetation is sparse, and few people live in this high altitude.

Section Wrap-Up

Answer these questions to check your understanding of the entire section.

1. **Organizing** Complete this chart by identifying the two tropical climate zones and five additional climate zones found in Latin America.

Tropical Climate Zones	Other Climate Zones

2. **Explaining** What factors contribute to the Atacama Desert being so dry?

_____
_____

In the space provided, write a paragraph explaining why most people live in the tierra templada *climate zone rather than in the other three altitude zones in the Andes.*

_____
_____
_____
_____
_____
_____
_____
_____
_____
_____
_____
_____

Chapter 8, Section 1 (Pages 208–215)

History and Governments

Big Idea

All living things are dependent upon one another and their surroundings for survival. As you read, complete the chart below. List key facts about the civilizations of the region.

	Key Facts
Olmec	
Maya	
Toltec	
Aztec	
Inca	

Notes | Read to Learn

Early History (pages 209–211)

Listing

List four achievements of the Maya.

1. _____
2. _____
3. _____
4. _____

Stating

How many people lived in Tenochtitlán?

_____

The Olmec built Latin America's first civilization in southern Mexico. It lasted from 1500 B.C. to 300 B.C. Some Olmec cities grew **maize,** or corn, as well as squash and beans. Some cities controlled mineral resources such as **jade,** a green semiprecious stone, and **obsidian,** a hard, black, volcanic glass used to make weapons. Other cities were religious centers.

The Maya lived in Mexico's Yucatán Peninsula from about A.D. 300 to A.D. 900. They built huge pyramid temples, used astronomy to develop a calendar, and had a number system based on 20. They also used **hieroglyphics,** a form of writing that uses signs and symbols, to record their history. Around A.D. 900, the Maya civilization mysteriously collapsed. Around the same time, the Toltec conquered northern Mexico. They controlled trade and held a monopoly on obsidian, giving them the most powerful weapons in the region.

The Aztec arrived around 1200. They adopted Toltec culture, conquered neighboring peoples, and took control of trade. Their capital, Tenochtitlán, was built on an island in a lake. About 250,000 people lived there. Roads and bridges linked the city to the mainland.

52   Chapter 8, Section 1

Early History (continued)

Determining Cause and Effect

How did Hernán Cortés defeat the Aztec?

_____
_____
_____
_____

The Inca were powerful in Peru during the 1400s. Their **empire,** or large territory with many different peoples under one ruler, stretched 2,500 miles along the Andes. Roads and suspension bridges linked all parts of the empire to Cuzco, the capital. The Inca also had military posts, irrigation systems, and a complex system of record keeping.

Spanish explorers reached Latin America in the late 1400s. In 1521 Hernán Cortés defeated the Aztec and their simple weapons with guns, cannons, and horses. Diseases also wiped out the Aztec and Inca. In 1532 Francisco Pizarro attacked and quickly conquered the Inca Empire. These conquests brought Spain and its new empire enormous wealth. Other European countries also seized different parts of the Americas and established colonies. Portugal took over Brazil. The French, British, and Dutch settled Caribbean areas.

The Europeans set up colonial governments, spread Christianity, and forced Native Americans to grow **cash crops,** or farm products grown for export. After many Native Americans died from disease, enslaved Africans were brought to work on plantations and in mines.

Forming New Nations (pages 212–215)

Summarizing

What conflicts arose after independence?

1. _____
2. _____
3. _____

Naming

Name three exports of Latin America in the late 1800s.

1. _____
2. _____
3. _____

Inspired by the American and French revolutions, the people of Latin America fought for their freedom. In Haiti, Toussant-Louverture led a revolt that overthrew French rule in 1804. Simón Bolívar fought against Spain and won freedom for Venezuela, Colombia, Ecuador, and Bolivia in 1819. Mexico gained independence in 1821. José de San Martín helped lead Chile and Peru to freedom. Brazil broke away peacefully from Portugal in the 1820s.

After winning independence, Latin America faced political and economic challenges. Slavery was ended, but conflicts occurred over the role of religion, boundary lines, and the huge gap between rich and poor. Strong leaders known as **caudillos** were supported by the upper class and often made it difficult for democracy and prosperity to grow.

In the late 1800s, Latin America's economy depended on agriculture and mining. Foreign companies moved in to control the export of bananas, sugar, coffee, copper, and oil. The United States also increased its political influence in the region. It fought Spain to gain Cuba in 1898. Puerto Rico came under U.S. control. The United States helped Panama win its independence from

Chapter 8, Section 1

Forming New Nations (continued)

Determining Cause and Effect

What resulted from Latin America's increasing debt?

_____
_____
_____

Colombia in 1903. In return, Panama allowed America to build the Panama Canal. U.S. troops entered Haiti, Nicaragua, and the Dominican Republic to protect American economic interests. Many Latin Americans feared that the United States would try to control them. In response, the United States established the Good Neighbor Policy in the 1930s, promising not to send military forces and to respect Latin American rights.

In the mid-1900s, Latin American leaders borrowed money from other countries to encourage industrial growth. The increasing debt weakened local economies, however. Many people lost jobs and faced rising prices. Dissatisfied groups in some countries rebelled against ruthless leaders. In Cuba, Fidel Castro led a successful revolt and set up a **communist state,** in which the government controls the economy and society. Civil wars raged in other countries, such as El Salvador.

Economic and political reforms in the 1980s strengthened many Latin American countries. However, challenges still exist in the region. These include rapid population growth, limited resources, the illegal drug trade, and the vast division between the wealthy and the poor. Leaders elected in the early 2000s promised significant changes.

Section Wrap-Up

Answer these questions to check your understanding of the entire section.

1. **Explaining** What is obsidian? How did it help make the Toltec powerful?

   _____
   _____

2. **Sequencing** Complete the time line below with key events and dates in the history of Latin America. Extend the line and add more boxes if necessary.

On a separate sheet of paper, write a paragraph explaining whether you think U.S. involvement in Latin America has helped or hurt the region.

Chapter 8, Section 2 (Pages 218–224)
Cultures and Lifestyles

Big Idea

The characteristics and movement of people impact physical and human systems. As you read, complete the diagram below. Add one or more facts to each of the outer boxes.

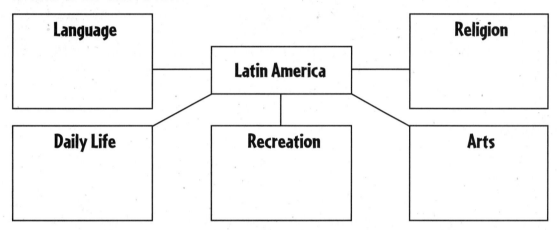

 **Read to Learn**

The People (pages 219–221)

Summarizing

Summarize three types of migration that occur in Latin America.

1. _____
2. _____
3. _____

Latin America has a high population growth rate. Central America has the fastest-growing populations. Guatemala and Honduras are expected to double their populations by the year 2050.

The region's climates and landscapes affect where people live. Areas with extreme temperatures, rain forests, deserts, and mountains are sparsely populated. Most people live in moderate climates along the coasts of South America or in an area stretching from Mexico into Central America. These areas have fertile soil and allow for easy movement of people and goods.

Latin America's population has been affected by **migration,** or the movement of people. People immigrate into the region in search of jobs or personal freedom. At the same time, some Latin Americans leave the region to escape political unrest or to find a better way of life. Many legally and illegally enter the United States looking for work. Other people move within the region, leaving their farms to search for jobs in rapidly growing cities.

Some of the world's largest cities are in Latin America, including Mexico City, São Paulo, Rio de Janeiro, and Buenos Aires. Millions of villagers that stream to these cities face poverty, crowded neighborhoods, lack of sanitation, and rising crime.

Chapter 8, Section 2   **55**

Notes | Read to Learn

The People (continued)

Locating

Where do most Mestizos live in this region?

_____
_____
_____

Defining

Define pidgin language and give an example of one.

_____
_____
_____

Ethnic groups in Latin America include Native Americans, Europeans, Africans, Asians, and people of mixed descent. Mexico, Central America, Ecuador, Peru, and Bolivia are home to most of the region's Native Americans. They try to maintain their languages and traditions while adopting features of other cultures.

Since the 1400s, millions of Spanish and Portuguese have settled in Latin America. Other Europeans immigrated as well. Argentina, Uruguay, and Chile are populated mainly by people of Spanish and Italian origin. The Caribbean islands and northeastern Brazil have large populations of African Latin Americans who are descendants of enslaved Africans. Large Asian populations are found in the Caribbean islands, Guyana, and Brazil.

These ethnic groups blended over the centuries. **Mestizos,** or people of mixed Native American and European descent, form the largest groups in Mexico, Honduras, El Salvador, and Colombia. People of mixed African and European backgrounds live in Cuba, the Dominican Republic, and Brazil.

Spanish is the most widely spoken language, although most Brazilians speak Portuguese. Quechua, spoken centuries ago by the Inca, is an official language of Peru and Bolivia. English and French are spoken on some Caribbean islands. Several countries developed a **pidgin language** by combining parts of different languages. Haiti's Creole, for example, is a mix of French and African languages.

Daily Life (pages 223–224)

Listing

What sports are popular in Latin America?

_____
_____

Christianity plays a significant role in Latin American cultures. Most people became Christians during colonial times. Other faiths include traditional Native American and African religions, Islam, Hinduism, Buddhism, and Judaism.

Family is central to the Latin American way of life. Multiple generations often live in the same house, and extended families tend to live near each other. The father is the leader and decision maker, although the mother is the leader in some parts of the Caribbean.

Sports are widely popular. Soccer is the primary sport, and baseball also has a strong following. Cuba was the second country to play baseball, after the United States. Many skilled baseball players have entered the U.S. professional leagues. Cricket is played in Caribbean countries that were once ruled by the British.

 **Read to Learn**

Daily Life (continued)

Discussing

List two holidays celebrated in Latin America and a key fact about each.

1. _____
   _____

2. _____
   _____

Religious and patriotic holidays are common. Many countries hold a large festival called **carnival** on the day before Lent begins. In Mexico, a holiday called Day of the Dead is celebrated in honor of family members who have died.

Foods blend the traditions of the region. Corn and beans are common in Mexico and Central America. Beans and rice make up the main diet of Caribbean islanders and Brazilians. Beef is the national dish in Argentina, Uruguay, and Chile.

Music and art also reflect the region's ethnic mix. For example, Cuban music uses African rhythms. In the 1930s, Mexican artists painted **murals,** or large paintings on walls, that were similar to the artistic traditions of the Maya and Aztec. Magic realism is a writing style invented by Latin American writers of the late 1900s. It combines fantastic events with the ordinary.

Section Wrap-Up *Answer these questions to check your understanding of the entire section.*

1. **Explaining** Why are Latin American cities growing so rapidly?

   _____

   _____

2. **Organizing** Complete this chart to show the regional distribution of ethnic groups.

Origins	Country(ies)
African descent	
Asian descent	
European descent	
Native American	

 *On a separate sheet of paper, write a paragraph describing a holiday you invent that can be celebrated with your family and friends. Identify the reasons for the holiday and why it is important to invite others.*

Chapter 8, Section 2

Chapter 9, Section 1 (Pages 232–236)
Mexico

Big Idea

Patterns of economic activities result in global interdependence. As you read, complete the chart below with key facts about Mexico's economic regions.

Region	Key Facts
North	
Central	

Read to Learn

Mexico's People, Government, and Culture (pages 233–234)

Summarizing

List three facts about Mexico's government.

1. _____
   _____
2. _____
   _____
3. _____
   _____

The population of Mexico is a mix of Spanish and Native American heritage. About two-thirds of the people are mestizos, and one-fourth are Native American. Rural traditions are strong, but 75 percent of the people live in cities. The capital and largest city, Mexico City, has nearly 22 million people. Reflecting Spanish culture, Mexico's cities are organized around public squares called **plazas.** These serve as centers of public life.

Mexico is a federal republic with power shared between the national and state governments. A strong national president serves only one six-year term. He or she has more power than the legislative and judicial branches. A revolution occurred in Mexico in the early 1900s. After that, one political party ruled the country for a long time. In the 1990s, people were frustrated by economic troubles and their lack of political power. In 2000 Mexicans elected a president from a different political party for the first time in 70 years.

Mexican culture has been influenced by Native Americans and Europeans. Folk arts such as wood carving reflect Native American traditions. European culture is seen in sports such as soccer. Carved and painted religious statues blend the cultures.

58 Chapter 9, Section 1

Mexico's People, Government, and Culture (continued)

Stating

What is a popular Mexican food?

_____

Artists and writers are national treasures. Diego Rivera and his wife, Frida Kahlo, were famous painters of the early 1900s. Famous authors include Carlos Fuentes and Octavio Paz, who wrote about the values of Mexico's people.

Mexicans hold celebrations called fiestas, which are highlighted with parades, fireworks, music, and dancing. Popular foods—in both Mexico and the United States—include tacos and enchiladas.

Mexico's Economy and Society (pages 234–236)

Comparing

What crops are grown in the North and South regions of Mexico?

North:

_____
_____
_____
_____

South:

_____
_____
_____
_____

Identifying

Which region is the most heavily populated?

_____

Mexico has a growing economy. Three distinct economic regions result from the country's physical geography and climate. The North has dry and rocky land. Farmers must use irrigation to grow cotton, grains, fruits, and vegetables for export. Grasslands support cattle ranches worked by cowhands called **vaqueros.** The North also has deposits of copper, zinc, iron, lead, and silver. Factories are located near the Mexico–United States border in cities such as Monterrey, Tijuana, and Cuidad Juárez. Many factories are **maquiladoras,** or foreign-owned plants that hire Mexican workers to assemble parts made in other countries. The finished products are then exported.

More than half of Mexico's people live in the Central region. It has a pleasant climate and fertile volcanic soil, allowing for productive farming. Workers in industrial cities such as Mexico City and Guadalajara make cars, clothing, household items, and electronic goods. The country's energy industry is centered along the Gulf of Mexico, near offshore oil and gas deposits.

Mexico's poorest economic region is the South. **Subsistence farms,** or small plots where farmers grow only enough food to feed their families, are common in the mountains. On coastal lowlands, wealthy farmers grow sugarcane and bananas on **plantations,** or large farms that raise a single cash crop. Coastal resorts such as Acapulco and Cancun attract tourists.

Mexico's economy is shifting in priority from agriculture to manufacturing. The North American Free Trade Agreement (NAFTA) helped Mexico increase trade with Canada and the United States. Factories there produce steel, cars, and consumer goods. Banking and tourism industries also contribute to the economy. Economic advances have raised the standard of living, especially in the North.

Chapter 9, Section 1

Notes | Read to Learn

Mexico's Economy and Society (continued)

Categorizing

Write a positive outcome and a negative outcome of Mexico's economic growth.

_____
_____
_____
_____

Economic growth also has raised concerns about the environment and dangers to the health and safety of workers. Pollution has increased, and smog often blankets Mexico City.

Mexico's population is growing rapidly. As people move to cities in search of jobs, many have crowded together in slums. Some Mexicans are **migrant workers** who travel to find work planting or harvesting crops. Migrant workers often cross the border into the United States, sometimes illegally, to work.

Many Native Americans are poor and live in rural areas. In the 1990s, Native Americans in southern Mexico rose up against the government and demanded changes to improve their lives. Their struggle has not been resolved.

Section Wrap-Up

Answer these questions to check your understanding of the entire section.

1. **Identifying** Provide three examples of Native American and European influences on Mexican culture.

   _____
   _____

2. **Describing** How has Mexico's economy changed in recent years?

   _____
   _____

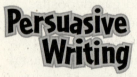

Choose one of the challenges facing Mexico. In the space provided, write a newspaper editorial in which you suggest steps Mexico's government could take to improve the situation.

_____
_____
_____
_____
_____

Chapter 9, Section 1

Chapter 9, Section 2 (Pages 237–240)
Central America and the Caribbean

Big Idea

The physical environment affects how people live. As you read, compare and contrast Guatemala and Costa Rica in the Venn diagram below.

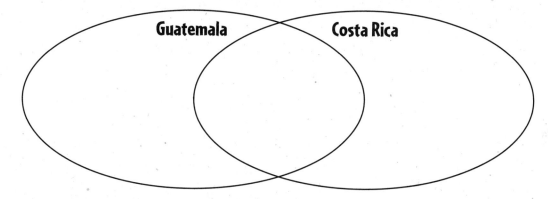

 **Read to Learn**

Countries of Central America (pages 238–239)

Specifying

Write down two economic changes in Guatemala in recent years.

1. _____
_____
2. _____
_____

Belize, Guatemala, El Salvador, Honduras, Nicaragua, Costa Rica, and Panama make up Central America. Most people in these countries farm. Bananas, sugarcane, and coffee are exported. Political and ethnic conflicts have weakened some economies in this region.

Guatemala

Half the people in Guatemala are descended from the ancient Maya. Many are of mixed Maya and Spanish heritage. Both languages are spoken in the country. A small but wealthy and powerful group owns most of the land in Guatemala. In the late 1990s, conflict erupted as rebel groups fought for control of the land. More than 200,000 people were killed or missing after the conflict ended.

Guatemala's economy has recently undergone some changes. Farmers are beginning to grow more valuable cash crops, such as fruits, flowers, and spices. In the early 2000s, Guatemala joined other Central American countries in a free trade agreement with the United States. This should allow the people to sell more goods to the United States.

Read to Learn

Countries of Central America (continued)

Differentiating

How does Costa Rica differ from its neighbors?

_____

_____

Costa Rica

Costa Rica has a stable democratic government. It has fought no wars since the 1800s. A police force—but no army—keeps law and order in the country. Costa Rica has fewer poor people than other countries in Central America. Costa Ricans have a high **literacy rate,** which is the percentage of people who can read and write. Literate workers generally earn higher wages because they are more productive.

Calculating

The Panama Canal was completed in 1914. For how long did the United States control it?

_____

Panama

Panama is located on the narrowest part of the isthmus of Central America. The United States built the Panama Canal there, which provides a shorter and faster route between the Atlantic and Pacific Oceans. The United States gave control of the canal to Panama in 1999. Panama earns money from the fees it charges shipping companies to use the canal. The canal area also attracts buyers and sellers. As a result, Panama has become a banking center.

Countries of the Caribbean (pages 239–240)

Stating

What happens to Cubans who criticize the government?

_____

_____

Some Caribbean countries, such as Cuba and Haiti, face political and economic challenges. Others, like Puerto Rico, are more stable.

Cuba

Cuba lies 90 miles south of Florida. It has a **command economy**—the communist government determines how resources are used and what goods and services are produced. The economy has not been successful, however, and most Cubans are poor. For many years, Cuba's economy relied on the sale of a single crop—sugar. The government is now trying to develop tourism and other industries to end the dependence on sugar.

Cuba's longtime dictator, Fidel Castro, controls society. People who criticize the government are often jailed. The United States condemns Cuba for these actions.

Haiti

Haiti is located on the western side of the island of Hispaniola. It has a troubled history. The government is unstable because of ongoing conflicts between political groups. Most Haitians live in poverty. A vital source of income is **remittances,** or money sent back home by Haitians who work in other countries.

Countries of the Caribbean (continued)

Summarizing

Summarize Puerto Rico's economy.

Products made:
_____
_____
_____

Crops grown:
_____
_____

Other:
_____

Puerto Rico

Since 1952, Puerto Rico has been a **commonwealth,** or a self-governing territory of the United States. The people are American citizens who can travel freely between Puerto Rico and the United States.

Puerto Rico has a higher standard of living than other countries in the Caribbean. Factories produce medicines, machinery, and clothing. Farmers grow sugarcane and coffee. The tourism industry also thrives in Puerto Rico.

Section Wrap-Up

Answer these questions to check your understanding of the entire section.

1. **Analyzing** What is the significance of the high literacy rate in Costa Rica?
_____
_____

2. **Explaining** Why is the Panama Canal important to the economy of Panama?
_____
_____

Expository Writing

In the space provided, write a paragraph explaining how Cuba's communist government has affected the Cuban economy and society.

_____
_____
_____

Chapter 9, Section 2

Chapter 9, Section 3 (Pages 246–252)
South America

Big Idea

People's actions can change the physical environment. As you read, describe Brazil's economy on the diagram below. Write the main idea on the single line and supporting details on the lines to the right.

Notes — Read to Learn

Brazil (pages 247–249)

Defining

What are favelas, and why have they emerged in Brazil?

_____
_____
_____

Identifying

What does Brazil use to make a substitute for gasoline?

_____

Brazil is the largest country in South America. Its culture is largely Portuguese rather than Spanish. Brazil's 187 million people are of European, African, Native American, Asian, and mixed ancestry. Most live in cities along the Atlantic coast, such as São Paulo and Rio de Janeiro. Millions have moved to coastal cities in search of jobs, settling in overcrowded slum areas called **favelas.** To reduce crowding, the government is encouraging people to move back to less-populated areas. The capital, Brasília, is located 600 miles inland.

Brazil is the world's leading producer of coffee, oranges, and cassava. Agricultural output has grown as more land has been cleared to grow crops. Machinery is used to perform many tasks, and crops have been scientifically changed to produce more and prevent disease.

Valuable minerals mined in Brazil include iron ore, bauxite, tin, manganese, gold, silver, and diamonds. Energy resources include offshore oil deposits and hydroelectric power. Sugarcane is used to make a substitute for gasoline. Booming industries produce machinery, airplanes, cars, food products, medicines, paper, and clothing.

64   Chapter 9, Section 3

Brazil (continued)

Listing

What economic activities occur in the selva?

_____

_____

_____

Brazil's greatest natural resource is the Amazon rain forest, called the **selva.** To promote economic development, the government has encouraged mining, logging, and farming in the rain forest. However, deforestation harms the rain forest's ecosystem and biodiversity, reduces the amount of oxygen released, and may affect Earth's climate patterns. Brazil has agreed to protect some rain forest areas.

The country is a democratic federal republic. Citizens elect the president and other leaders. Brazil's national government is much stronger than its 26 state governments.

Argentina (pages 249–250)

Paraphrasing

Complete these sentences.

Argentina had a high

_____

because it borrowed money from

_____.

Argentina was unable to make all of its

_____,

so it had to

_____

on its debts.

Argentina is the second-largest country in South America. The Andes tower in the west. Central Argentina has vast grasslands known as the Pampas. Most people have Spanish and Italian origins, and more than one-third live in the beautiful capital, Buenos Aires. This bustling port has been nicknamed "the Paris of the South."

Farming and ranching are vital to Argentina's economy. Cowhands called **gauchos** tend livestock on the Pampas. Gauchos, the national symbol of Argentina, are admired for their independence and horse-riding skills. Beef and beef products are the country's main exports. Argentina's factories produce food products, cars, chemicals, and textiles. Zinc, iron ore, copper, and oil are mined in the Andes.

In the 1990s, Argentina borrowed money from foreign banks, leading to a high **national debt,** or money owed by the government. Argentina had to **default** on its debts, meaning that it missed debt payments to the banks that lent the money. The economy has since recovered, and part of the debt has been repaid.

After gaining independence in the early 1800s, Argentina was led by military leaders. Today Argentina is a democratic federal republic. A powerful president is elected every four years.

Other Countries of South America (pages 251–252)

Venezuela, located along the Caribbean Sea, is a leading producer of oil and natural gas. Factories make steel, chemicals, and food items. Farmers grow sugarcane and bananas or raise cattle. Yet many Venezuelans are poor. Some live in slums

Chapter 9, Section 3

Other Countries of South America (continued)

Specifying
What was supposed to improve the lives of poor Venezuelans?

_____

Stating
What are the key elements of Venezuela and Chile's economies?

Venezuela:

_____

_____

Chile:

_____

surrounding Caracas, the capital. In 1998 Venezuelans elected Hugo Chávez as president. A former military leader, he promised to use oil money to improve the lives of the poor. However, his strong rule has divided the country.

Colombia borders both the Caribbean Sea and the Pacific Ocean. Most people live in the valleys and plateaus of the Andes. Bogotá, the capital, is located on an Andean plateau. The country mines coal, oil, and copper and is the world's leading supplier of emeralds. It exports bananas, sugarcane, rice, cotton, and world-famous coffee. Despite its economic strengths, Colombia faces much unrest. Wealth is in the hands of a few, while many people are poor. Drug dealers pose another problem. They pay farmers to grow coca leaves to produce cocaine.

Long, ribbon-shaped Chile borders the Pacific Ocean. Its landscapes vary from dry deserts in the north, to central fertile valleys, to glaciers in the south. Chile's economy is based on mining copper, gold, silver, iron ore, and **sodium nitrate,** a mineral used in fertilizer and explosives. In addition, farmers raise wheat, corn, beans, sugarcane, potatoes, grapes, and apples. People also raise cattle and sheep or work in the large fishing industry. Like Argentina, Chile was under military rule for many years. It is now a democracy.

Section Wrap-Up
Answer these questions to check your understanding of the entire section.

1. **Explaining** Why has Brazil's agricultural output increased?

_____

_____

2. **Comparing** Complete this chart by listing the types of governments.

Country	Type of Government
Brazil	
Argentina	
Chile	

On a separate sheet of paper, write a paragraph explaining why people in other areas of the world are concerned when Brazilians harm their rain forest.

Chapter 10, Section 1 (Pages 274–280)
Physical Features

Big Idea

Geographic factors influence where people settle. As you read, complete the chart below with key facts about the landforms, waterways, and resources of Europe.

Features	Key Facts
Landforms	
Waterways	
Resources	

Notes — Read to Learn

Landforms and Waterways (pages 275–277)

What are five types of landforms found in Europe?

1. _____
2. _____
3. _____
4. _____
5. _____

The continent of Europe shares a common landmass with Asia. This landmass is called Eurasia. Europe is located on the western portion of Eurasia.

Europe has a long coastline. It borders the Atlantic Ocean and the Baltic, North, Mediterranean, and Black Seas. Only a few countries are **landlocked,** or do not border an ocean or a sea. Nearness to water has influenced Europe's history and people. Shipping and fishing encouraged trade and helped build Europe's economy. Exploration spread European culture worldwide and brought ideas from Asia, Africa, and the Americas to Europe.

Peninsulas and Islands

Europe is a huge peninsula. It has many smaller peninsulas branching out from it. Europe also has many islands, including Great Britain, Ireland, Iceland, and Cyprus. At one time, seas, rivers, and mountains separated people living on these peninsulas and islands. Thus, many different cultures developed.

Read to Learn

Landforms and Waterways (continued)

Defining

What does the word navigable *mean?*

_____
_____
_____
_____
_____

Plains

The Northern European Plain is Europe's major landform. The soil is rich, and the plain also holds underground deposits of coal, iron ore, and other minerals. Most of Europe's population live and work on this vast plain. Other lowland areas include the Hungarian Plain and Ukrainian Steppe.

Mountains and Highlands

Europe's highest mountain ranges form the Alpine Mountain System. The Alps, the Pyrenees, and the Carpathians are included in this system. People and goods travel through **passes,** or low areas between mountains. Several other highland areas are used for mining and grazing livestock.

Waterways

Many rivers, lakes, and other waterways are found in Europe. The Danube and Rhine are two of Europe's longest rivers. Many of the rivers are **navigable**—wide and deep enough for ships to use. People and goods travel on the rivers throughout Europe and to the open sea. Fast-flowing rivers also provide electricity.

Europe's Resources (pages 278–279)

Identifying

Identify 10 of Europe's natural resources.

1. _____
2. _____
3. _____
4. _____
5. _____
6. _____
7. _____
8. _____
9. _____
10. _____

Europe has many valuable natural resources. These resources have helped Europe become a leader in the world economy.

Energy Resources

Coal has been a key energy source for Europe. In the 1800s, coal fueled early factories. Today Europe supplies almost half of the world's coal. Many people in Europe work as coal miners.

Two other energy resources are natural gas and petroleum. Productive oil fields are found beneath the North Sea in areas controlled by Norway and the United Kingdom.

Europeans also use clean energy sources that cause less pollution. Swift-flowing rivers create hydroelectric power. Wind farms use turbines with fanlike blades to make electricity.

Other Natural Resources

Other resources include iron ore and manganese used to make steel. Marble and granite provide building materials. Many of Europe's once-vast forests have been cut down, however.

Fertile soil allows farmers to grow large amounts of crops, including rye, oats, wheat, and potatoes. Europe's waterways provide another valuable resource—fish.

Environmental Issues (pages 279–280)

Problem-Solving

Describe how Europeans are attempting to solve each problem below.

1. Air pollution, acid rain
   _____
   _____
   _____

2. Water pollution
   _____
   _____
   _____

Air Pollution and Acid Rain

Smoke from burning oil and coal creates air pollution, which causes breathing problems and other health risks. When chemicals in air pollution mix with precipitation, acid rain results. Acid rain harms trees and damages the surfaces of buildings. Acid also builds up in lakes and rivers, poisoning fish and other wildlife.

Water Pollution

The waterways in and around Europe are polluted. Sewage, garbage, and industrial waste are dumped into the region's seas, rivers, and lakes. Chemicals in pesticides and fertilizers run off from farmland into rivers, harming fish and other marine life.

Finding Solutions

Europeans are working to fix environmental problems. Factories are trying to release fewer chemicals into the air. Lakes are being treated with lime to reduce acid rain damage. Waste and sewage are being treated to provide cleaner water. Farmers are using less fertilizer to reduce the amount of chemical runoff. More Europeans are recycling and reducing the amount of garbage they produce.

Section Wrap-Up — *Answer these questions to check your understanding of the entire section.*

1. **Evaluating** How are waterways important to Europe?
   _____
   _____

2. **Determining Cause and Effect** How have Europe's resources contributed to environmental problems?
   _____
   _____

On a separate sheet of paper, write a paragraph comparing and contrasting the landforms in your area to the landforms found in Europe. Which are similar? Which are different?

Chapter 10, Section 1

Chapter 10, Section 2 (Pages 282–288)
Climate Regions

Big Idea

The physical environment affects how people live. As you read, complete the chart below by listing and describing Europe's main climate zones.

Climate Zone	Characteristics

Notes — Read to Learn

Wind and Water (page 283)

Finding the Main Idea

What is the main idea of this subsection?

_____
_____
_____
_____

Europe is farther north than much of the United States. Because of this, you might expect its climate to be cold. However, the North Atlantic Current carries warm water from the Gulf of Mexico to Europe. Winds from the west, called westerlies, pass over this current and also carry warmth to Europe.

Other wind patterns affect regions in Europe. Winds blowing north from Africa warm southern Europe. In contrast, winter winds from Asia can lower temperatures in eastern Europe.

The waters around Europe also influence its climate. Winds blowing from the seas help cool the land in summer. In winter, the same winds warm the cold land. Thus, coastal Europe has a more moderate climate than inland areas.

Climate Zones (pages 284–288)

Europe has three main climate zones—marine west coast, humid continental, and Mediterranean. Five additional climate zones appear in small areas of Europe—subarctic, tundra, highland, steppe, and humid subtropical.

Read to Learn

Climate Zones (continued)

Identifying
Underline the types of vegetation for each climate zone.

Explaining
How do mountains affect southern Europe's climate?

_____
_____
_____
_____
_____

Differentiating
Complete this sentence.

_____ *winds are cold and dry, and* _____ *winds are hot and dry.*

Marine West Coast

Much of northwestern and central Europe has a marine west coast climate. This climate features mild temperatures and much rain. The climate supports a long growing season. Rainfall occurs mostly in the autumn and early winter. Forests flourish in the marine west coast climate. Some forests have **deciduous** trees, which lose their leaves in the fall. In cooler areas of the zone, **coniferous** trees, also called evergreens, grow.

Humid Continental

The second main zone is the humid continental climate. It is found in eastern Europe and parts of northern Europe. This zone has cooler summers and colder winters than the marine west coast zone. It also gets less rain and snow. The humid continental zone has mixed forests of deciduous and coniferous trees.

Mediterranean

The Mediterranean zone is Europe's third major climate zone, and it includes much of southern Europe. Mediterranean summers are hot and dry, and winters are mild and wet. The Pyrenees and Alps stop cold winds from blowing into southern Europe. Southern France, however, experiences cold, dry, **mistral** winds from the north.

Hot, dry winds from Africa, known as **siroccos,** pick up moisture as they cross the Mediterranean Sea. They bring humid weather to much of southern Europe. Vegetation in the Mediterranean zone includes olive trees, grapes, and low-lying shrubs—plants that do not need much water.

Subarctic and Tundra

The cold subarctic zone covers parts of Norway, Sweden, and Finland. Evergreen trees grow in this zone. Even farther north is the tundra zone, a frigid area of treeless plains. Only low shrubs and mosses can grow in the tundra. During winter, the sun's rays reach this region for only four hours per day.

Highland

The highland climate zone is found in the high altitudes of the Alps and Carpathian Mountains. This climate is generally cool to cold. Temperature and rainfall vary in this zone, depending mainly on a place's altitude, or elevation above sea level. Evergreens grow part of the way up the mountains, stopping at a point called the timberline. Only scrubby bushes and low-lying plants grow above the timberline.

Notes | Read to Learn

Climate Zones (continued)

Listing
What are the five smaller climate zones in Europe?

1. _____
2. _____
3. _____
4. _____
5. _____

Other Climate Zones
The southern part of Ukraine is a dry, treeless grassland called a steppe. The climate—also called steppe—is dry too, but not as dry as a desert. An area of land north of the Adriatic Sea has a humid subtropical climate. Summers are hot and wet here, and winters are mild and wet.

Climate Change
Many European leaders are worried about global warming. They fear that melting glaciers will cause ocean levels to rise, flooding coastal areas of Europe. Millions of Europeans live on or near the coasts. Officials are encouraging people to change their patterns of energy use. Most European governments have also signed the Kyoto Treaty, which limits the output of greenhouse gases that cause global warming.

Section Wrap-Up
Answer these questions to check your understanding of the entire section.

1. **Defining** What are westerlies? How do they affect the climate of Europe?

   _____
   _____

2. **Specifying** In which of Europe's climate zones do coniferous, or evergreen, trees grow?

   _____

Descriptive Writing
Imagine that you live in the tundra climate zone. Write a paragraph describing what a typical winter day might be like.

_____
_____
_____
_____
_____

Chapter 11, Section 1 (Pages 294–303)
History and Government

Big Idea

The characteristics and movement of people impact physical and human systems. As you read, complete the time line below with at least five key events and dates in Europe's history.

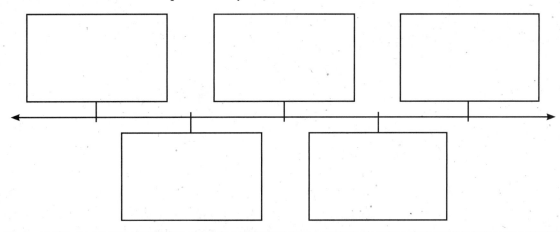

Notes	Read to Learn

Ancient Europe (pages 295–297)

How did ancient Greece and Rome influence later civilizations?

Ancient Greece:

_____

_____

_____

Ancient Rome:

_____

_____

_____

European civilizations began about 2,500 years ago in ancient Greece and Rome—known as the **classical** world. Greece's many mountains, islands, and seas separated early communities. Groups formed independent **city-states,** made up of the city and the area surrounding it. The city-state of Athens became the world's first **democracy,** in which citizens shared in running the government. Athens's democracy set an example for later civilizations. Learning and the arts also thrived.

Philip II of Macedonia conquered Greece in the mid-300s B.C. His son, Alexander the Great, made more conquests, spreading Greek culture to Egypt, Persia, and India. Rome conquered Greece in about 130 B.C.

Rome, located on the Italian Peninsula, became a **republic** in 509 B.C. In a republic, citizens choose their leaders. Rome also developed a code of laws known as the Twelve Tables. About 200 B.C., Roman armies began to conquer lands in the Mediterranean region. The republic grew into the massive Roman Empire. In A.D. 27, Augustus became Rome's first **emperor,** or all-powerful ruler. Christianity arose during his rule, and in A.D. 392, it became Rome's official religion.

Notes | Read to Learn

Ancient Europe (continued)

Attacks from Germanic groups finally caused the Roman Empire to fall in A.D. 476. But Rome's influence was long lasting. Roman law shaped later legal systems. The language of Rome was Latin, the basis of many modern languages. Rome's architectural style became a model for Europe and the West.

Expansion of Europe (pages 298–300)

Identifying

1. What were the two Christian branches in Europe during the Middle Ages?

   _____
   _____

2. Which two religious groups fought in the Crusades?

   _____
   _____

Making Connections

Underline the phrase that explains why feudalism began. Then circle the phrase that explains why feudalism declined.

After Rome's fall, Europe entered a 1,000-year period known as the Middle Ages. Life in western Europe revolved around the Roman Catholic Church, led by the **pope.** In eastern Europe, the Byzantine Empire spread Christianity through the Eastern Orthodox Church.

Charlemagne, a powerful king, united much of western Europe in the A.D. 800s. After he died, his empire broke into many small, weak kingdoms. To protect their power, kings gave land to nobles. In return, the nobles served as the king's army in a system known as **feudalism.**

Christianity united much of Europe. But Muslims, or followers of Islam, spread through Southwest Asia and North Africa after the A.D. 600s. They also seized Palestine, which Christians viewed as their Holy Land. This led to religious wars called the Crusades. The Crusades opened trade with Muslim lands, which provided tax money for kings. Feudalism declined. Europe's kingdoms began to develop into **nation-states**—countries whose people shared a common culture or history.

The Renaissance, or "rebirth," was a time of renewed interest in art and learning from 1350 to 1550. The Renaissance thrived in Italian city-states like Florence, Rome, and Venice. Artists such as Michelangelo and Leonardo da Vinci created masterpieces. Humanism, or the belief in the importance of the individual rather than the Church, began to emerge.

In 1517 a religious leader named Martin Luther tried to reform, or change, some church practices he thought were wrong. The Reformation led to Protestantism, a new form of Christianity. Religious wars erupted throughout Europe. Monarchs gained power as church leaders lost theirs.

With new power and new technology, monarchs sent explorers overseas to look for spices and gold. Spain grew wealthy after Christopher Columbus landed in the Americas in 1492. Soon European monarchs began setting up colonies in the Americas, Asia, and Africa.

Modern Europe (pages 301–303)

Specifying

Underline and number three "revolutions" that swept Europe beginning in the 1600s.

Summarizing

What is the goal of the European Union?

_____

_____

Europe experienced several **revolutions,** or sweeping changes, beginning in the 1600s. During the Scientific Revolution, people used science and reason as a guide, rather than faith or tradition. The 1700s became known as the Age of Enlightenment. Thinkers such as John Locke said that all people have natural rights, including the rights to life, liberty, and property. Political revolutions in several countries limited the power of government.

During the Industrial Revolution, which began in Britain, machines and factories made goods faster and cheaper. Many Europeans left their farms to find work in cities.

Industry helped countries grow more powerful. They developed new weapons and competed for colonies. Tensions soon led to World War I (1914–1918) and World War II (1939–1945). Europe was left in ruins, and millions of people died. Six million Jews were killed in the **Holocaust.**

After World War II, the United States and Soviet Union fought for world power during the Cold War era. Much of Western Europe allied with the United States. Eastern European countries allied with the Communist Soviet Union. **Communism** is a system in which government controls the ways of producing goods. After the Soviet Union broke apart in 1991, many former Communist countries became democratic.

In 1993 several countries formed the European Union. Other countries have since joined. Their goal is to unify Europe. Workers and products move freely among member countries.

Section Wrap-Up *Answer these questions to check your understanding of the entire section.*

1. **Explaining** What role did religion play in the history of Europe?

_____

_____

2. **Listing** What important changes occurred during the Middle Ages?

_____

_____

On a separate sheet of paper, write a paragraph describing one or more of the basic rights you have as an American citizen. Include ideas from the Enlightenment in your paragraph.

Chapter 11, Section 1

Chapter 11, Section 2 (Pages 306–312)
Cultures and Lifestyles

Big Idea

Europe is home to many different cultural groups. As you read, complete the diagram below by listing five key facts about Europe's population patterns.

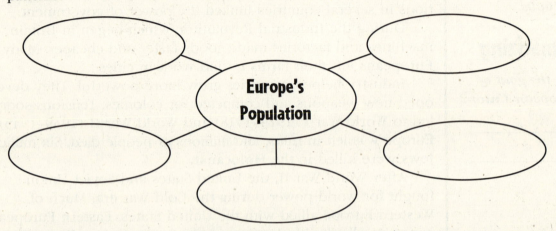

Notes | Read to Learn

Population Patterns (pages 307–308)

Determining Cause and Effect

What has caused Europe's ethnic mix?

_____
_____
_____

A Rich Ethnic Mix

Many different ethnic groups live in Europe. An **ethnic group** is a collection of people who share the same ancestry, language, and customs. Migrations, wars, and changing boundaries have led to Europe's ethnic mix. Ethnic loyalties create bonds. They also create conflict. Fighting among ethnic groups resulted in Yugoslavia splitting into five separate countries in 1990. Europeans are working toward unity, however. They value democracy and human rights. They expect their governments to care for citizens. Many European countries are **welfare states,** in which the government provides care for the sick, needy, and retired.

Population Changes

Europe's population continues to change. Since World War II, people have immigrated from Asia, Africa, and Latin America. Immigrants are not always welcomed. They compete with residents for jobs, housing, and other services. Some countries help immigrants adapt quickly to their new nation. Others pass laws to prevent immigrants from coming in at all.

Population Patterns (continued)

The continent's total population is declining. The reason is that Europe's **fertility rate**—the average number of children born to each woman—is low. Experts predict that by the year 2050, Europe will have 10 percent fewer people. Fewer people will mean fewer workers to keep the economy growing. In addition, young workers will have to pay higher taxes to support an increasingly older population.

Life in Europe (pages 308–310)

Listing
What are four ways to travel in Europe?

1. _____
2. _____
3. _____
4. _____

Explaining
Underline the sentence that explains why people in eastern European countries have lower incomes than those in western European countries.

Cities and Transportation

The Industrial Revolution in the late 1700s changed Europe from a rural, farming society to an urban, industrial one. **Urbanization** resulted, with most people living in cities and towns. Today Paris and London are two of the largest cities in the world.

Most European cities have government-owned public transportation systems. A vast rail system links cities and towns throughout Europe. France developed high-speed trains, which help protect the environment. Subways are common. Trains even speed through an underwater tunnel—called the Chunnel—between England and France. Highways also allow high-speed travel, particularly Germany's autobahn. Canals and rivers are widely used to ship goods, and ports dot Europe's long coastline. Airports also connect European cities.

Education, Income, and Leisure

Europeans are well educated and have a high literacy rate. As a result, many Europeans earn more money than people in other parts of the world. Service industries provide many jobs. However, incomes in northern and western Europe are higher than those in southern and eastern Europe. Many eastern European countries are still rebuilding from ethnic conflicts or from years under Communist rule.

The overall higher incomes mean that people have more money to spend on leisure activities. Europeans like to travel. France and Italy are popular vacation spots. Sports are popular, too. Ice hockey, skiing, rugby, and soccer are favorite pastimes.

Chapter 11, Section 2

Notes | Read to Learn

Religion and the Arts (pages 310–312)

Sequencing

How did European art change from ancient times to the 1900s?

_____
_____
_____
_____
_____
_____
_____

Many Europeans are **secular,** or nonreligious. Yet religion has had a major impact on the life and art of Europe. Roman Catholicism is practiced in western Europe and some eastern European countries. Northern Europeans are mainly Protestant. Eastern Orthodoxy is practiced in the southern part of eastern Europe. Judaism and Islam have also influenced Europe's culture. For the most part, Europeans of different religions live together peacefully. But religious differences in Europe have sometimes led to violence.

Arts

The ancient Greeks and Romans built temples with huge, graceful columns. Gothic cathedrals with pointed arches and stained-glass windows arose during the Middle Ages. Art often focused on holy subjects or religious symbols. Renaissance artists and writers focused on religion too, but they also portrayed lifelike figures and believable characters in their works. In the 1800s, artists, writers, and musicians tried to stir emotions in a style known as Romanticism. Impressionists used bold colors to create "impressions" of the natural world. In the 1900s, painters expressed feelings and ideas in abstract paintings.

Section Wrap-Up

Answer these questions to check your understanding of the entire section.

1. **Explaining** How do individual European countries deal with immigration?

   _____
   _____
   _____

2. **Drawing Conclusions** Why do most Europeans have higher incomes than people in other parts of the world?

   _____
   _____
   _____

On a separate sheet of paper, write a paragraph explaining some of the problems that Europeans will face as their population declines.

Chapter 12, Section 1 (Pages 320–328)
Northern Europe

Big Idea

Geographers organize the Earth into regions that share common characteristics. As you read, complete the chart below. List key facts about the people and cultures of northern Europe.

People and Cultures		
United Kingdom	**Ireland**	**Scandinavia**

Notes — Read to Learn

The United Kingdom (pages 321–324)

Naming

What four regions make up the United Kingdom?

1. _____
2. _____
3. _____
4. _____

Identifying

What waterway separates Great Britain from mainland Europe?

_____

The United Kingdom is an island nation northwest of mainland Europe. England, Scotland, and Wales make up the island of Great Britain. Northern Ireland, also part of the United Kingdom, is located in one corner of the nearby island of Ireland. The English Channel separates Great Britain from the rest of Europe. Britain's southern and eastern plains have fertile farmland. The highlands and mountains of Scotland and Wales are best suited to sheep herding. London, the capital, is a world center of finance and business.

Great Britain led the Industrial Revolution, and it still exports manufactured goods. Electronics industries, banking, and health care also are vital to the economy. Oil and natural gas beneath the North Sea supply most of Britain's energy.

The United Kingdom is both a **constitutional monarchy** and a **parliamentary democracy.** A king or queen is the ceremonial head of state, but voters elect members of Parliament, the lawmaking body. The leader of the political party with the most members in Parliament becomes prime minister. Scotland, Wales, and Northern Ireland have regional legislatures with control over education and health care.

Chapter 12, Section 1   79

Notes | Read to Learn

The United Kingdom (continued)

Specifying

Circle the three languages spoken in Great Britain.

Great Britain has the third-largest population in Europe. Almost 90 percent of the people live in cities.

British people speak English. Welsh and Scottish Gaelic also are spoken in some areas. Most of the people are Protestant Christians, but a growing number of immigrants practice Islam, Hinduism, and Sikhism.

The United Kingdom was a powerful empire in the 1700s and 1800s. British culture—including the English language, the sport of cricket, and British literature—spread to many lands.

The Republic of Ireland (pages 325–326)

Defining

Define the word productivity by using it in a sentence.

_____
_____
_____
_____

The Republic of Ireland is an independent Catholic country. Nicknamed the Emerald Isle, Ireland is lush and green because of its regular rainfall. **Peat,** or plants partly decayed in water, is found in low-lying areas. The peat is dug from **bogs**—swampy lands. The peat is dried and can be burned for fuel.

Farmers raise sheep and cattle and grow sugar beets and potatoes. More people work in manufacturing than farming. Ireland's industries produce clothing, pharmaceuticals, and computer equipment. Increased **productivity**—how much work a person does in a set amount of time—has boosted the economy.

Celts settled the island hundreds of years ago. Ireland's two languages are Irish Gaelic (a Celtic language) and English. Dublin is the capital. Irish music and folk dancing are performed all over the world. Irish playwrights, novelists, and poets had a great influence on world literature.

Catholics in Northern Ireland would like to unite with their southern neighbors. Protestants there want to remain part of the United Kingdom, however. This dispute led to violence from the 1960s to the 1990s. A 1998 agreement has not halted the dispute.

Scandinavia (pages 326–328)

Naming

Underline the five countries that make up Scandinavia.

Scandinavia is the northernmost region of Europe. It includes five countries—Norway, Sweden, Finland, Denmark, and Iceland. Their standards of living are among the highest in the world.

The northernmost part of Scandinavia is always cold. The southern and western areas are mild due to the warm North Atlantic Current. Islands dot the jagged coastline. Lowland plains cover Denmark and southern Sweden and Finland. Mountains rise

Notes | Read to Learn

Scandinavia (continued)

Listing

List seven features of Scandinavia's landscape.

1. _____
2. _____
3. _____
4. _____
5. _____
6. _____
7. _____

along Norway and Sweden's shared border. Forests and lakes cover Sweden and Finland. Barren, frozen tundra is found above the Arctic Circle. In Iceland, tectonic activity creates springs called **geysers** that shoot hot water and steam into the air. Norway has many narrow sea inlets called **fjords**, which are surrounded by steep cliffs carved by glaciers.

The strong economies of Scandinavia consist of agriculture, manufacturing, and service industries. Fishing also is important. Each country uses different sources of energy. Norway relies on oil and natural gas pumped from its fields under the North Sea. Iceland uses **geothermal energy**, or electricity produced by underground steam. Finland generates hydroelectric power from its fast-running rivers. Sweden uses nuclear power and oil.

Norway, Sweden, Denmark, and Iceland share ethnic ties and speak similar languages. Finland's language and culture are different, yet it shares close historic and religious links to the other countries. Most people are Protestant Lutherans. Denmark, Norway, and Sweden are constitutional monarchies like the United Kingdom. Finland and Iceland are republics with elected presidents. All five countries are welfare states. They provide health care, child care, elder care, and retirement benefits to all. In return for these services, citizens pay some of the highest taxes in the world.

Section Wrap-Up

Answer these questions to check your understanding of the entire section.

1. **Differentiating** What are some of the differences between Northern Ireland and the Republic of Ireland?

   _____
   _____

2. **Explaining** What is a welfare state?

   _____
   _____

Expository Writing

On a separate sheet of paper, write a paragraph explaining how the government of the United Kingdom is different from the government of the United States.

Chapter 12, Section 1

Chapter 12, Section 2 (Pages 329–337)
Europe's Heartland

Big Idea

People's actions can change the physical environment. As you read, complete the Venn diagram below. Compare and contrast France and Germany.

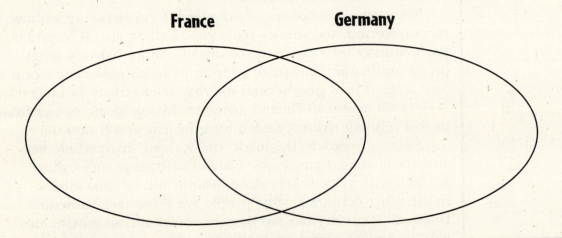

France   Germany

Notes | Read to Learn

France and the Benelux Countries (pages 330–333)

Specifying

What products do French farmers specialize in?

_____

_____

Stating

What are two landmarks in Paris?

_____

_____

France lies in the western part of Europe. Its mild climate and rich soil are good for farming. French farms are known for their **specialization,** or focusing on using their best resources. Some farms grow grapes for wine, while other farms produce cheese. Traditional industries make cars and trucks, chemicals, and textiles. **High-technology industries** make computers and other products that require specialized engineering. Tourism is another major part of the French economy. Millions of tourists visit Paris, the capital, and its museums and cathedrals, such as the Louvre and Notre Dame. They also tour the country's historic castles.

Most people speak French and are Roman Catholic. Immigration from Muslim countries has made Islam France's second-largest religion. The majority of people live in urban areas.

The French are proud of their culture. French cooking and fashion are admired worldwide. France also boasts famous writers, philosophers, artists, and composers. France is a democratic republic with an elected president and a prime minister appointed by the president.

France and the Benelux Countries (continued)

Naming
What are the three Benelux countries?

_____
_____
_____

Explaining
How have the Dutch increased their farmland?

_____
_____
_____

The Benelux Countries

Belgium, Netherlands, and Luxembourg—also known as the Benelux countries—have a low, flat landscape. Most people live in crowded cities, work in businesses or factories, and have a high standard of living. All three countries are parliamentary democracies with monarchs.

Belgium has three distinct cultural regions: Flanders has Dutch-speaking Belgians, Wallonia has French-speaking Belgians, and the Brussels region is **bilingual,** or has two official languages. Brussels is the capital and headquarters of the European Union (EU).

The people of the Netherlands are known as the Dutch. They have built dikes and drained land that once was covered by the sea, turning it into rich farmland called **polders.** The Dutch make good use of their limited space, building narrow but tall homes. Amsterdam is the capital and largest city.

Luxembourg, a center of trade and finance, is the headquarters for many **multinational companies,** or firms that do business in several countries. The people have a mixed French and German background.

Germany and the Alpine Countries (pages 334–337)

Listing
List four things that Germany produces.

1. _____
2. _____
3. _____
4. _____

Germany has flat plains in the north, rocky highlands in the center, and the Alps in the far south. The Danube, Elbe, and Rhine Rivers are used to transport raw materials to factories and finished goods to markets.

Separate states made up the region for centuries. In 1871 these states joined to create the country of Germany. In the early 1900s, Germany's attempts to control Europe led to two world wars. In 1945 the Soviet Union controlled Communist East Germany. West Germany became democratic. **Reunification** came about in 1990, when the two parts of Germany were united into one country. Today Germany is a federal republic. A chancellor chosen by parliament is head of the government.

Germany has the largest population in Europe. Berlin is Germany's capital and largest city. Nearly everyone speaks German and is Protestant or Catholic. Germany is a global economic power and a leader in the European Union. The country grows enough food to feed its people and to export to other countries. Industry drives the strong economy, however. Factories produce steel, chemicals, cars, and electrical equipment.

Chapter 12, Section 2

 **Notes** | **Read to Learn**

Germany and the Alpine Countries (continued)

Identifying

What are Switzerland's four official languages?

1. _____
2. _____
3. _____
4. _____

Speculating

Why do you think people consider Swiss banks secure?

_____
_____
_____

The Alpine Countries

The Alps are a mountain range in central Europe. The countries in this region—Switzerland, Austria, and Liechtenstein—are known as the Alpine countries. Liechtenstein is the smallest, with a population of about 40,000.

Landlocked Switzerland practices **neutrality,** meaning it does not take sides in wars. Its stable democracy has made Switzerland a home to many international organizations, such as the International Red Cross. The rugged Alps separated Switzerland's people, resulting in diverse traditions, ethnic groups, and religions. The four official languages of Switzerland are German, French, Italian, and Romansch. Many Swiss speak more than one language.

Switzerland is a thriving industrial nation. Hydroelectricity powers its industries and homes. Swiss workers make electrical equipment, chemicals, watches, chocolate, and cheeses. People around the world consider Swiss banks safe and secure, so financial services make up a large part of Switzerland's economy.

Austria is a landlocked country east of Switzerland. The Alps cover most of the country and provide timber, iron ore, and beautiful scenery for tourists. Hydroelectricity powers Austria's factories, which produce machinery, chemicals, metals, and vehicles.

Most Austrians live in cities and towns, speak German, and are Roman Catholic. The capital, Vienna, is a center of culture. The city is known for its concert halls, palaces, and churches.

Section Wrap-Up

Answer these questions to check your understanding of the entire section.

1. **Sequencing** List three key events and dates in Germany's recent history on the time line below.

   |_____|_____|_____|

2. **Explaining** What are the Alpine countries, and why are they called that?

   _____
   _____

 **Persuasive Writing**

Do you think it is important for a person to know more than one language? On a separate sheet of paper, write a paragraph presenting your opinion and persuading your readers to agree.

84   Chapter 12, Section 2

Chapter 12, Section 3 (Pages 338–342)
Southern Europe

Big Idea

Places reflect the relationship between humans and the physical environment. As you read, complete the diagram below. Write three characteristics shared by Spain, Italy, and Greece

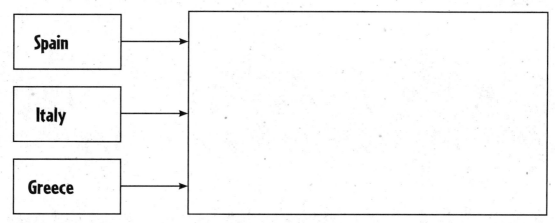

Notes | Read to Learn

Spain and Portugal (pages 339–340)

Specifying

What are Spain and Portugal's chief agricultural products?

Spain:

_____
_____
_____
_____

Portugal:

_____
_____

Spain and Portugal are located on the Iberian Peninsula. A third country, tiny Andorra, is nestled in the Pyrenees Mountains. The Meseta, a dry plateau surrounded by mountains, covers most of Spain. Because little rain falls on the Meseta, farmers use a technique called **dry farming** to grow wheat and vegetables. The land is left unplanted every few years so it can store moisture. The milder climate of southern Spain is good for growing citrus fruits, olives, and grapes.

Manufacturing and service industries dominate the economy. Spain produces foods, clothing, footwear, steel, and cars. Tourism also boosts the economy. Tourists visit Spain's castles, cathedrals, and Mediterranean beaches. They also enjoy Spanish bullfighting and flamenco dancing.

Castilian Spanish is Spain's official language, but people speak different languages in the country's various regions. The people of Catalonia speak Catalon. The Basques speak Euskera, a language unrelated to any other language in the world. Spain's democratic government has given the different regions a great deal of **autonomy,** or self-rule. In the Basque region, many people want to be independent.

Spain and Portugal (continued)

Locating

Where do most Portuguese live?

_____
_____
_____
_____

Spain's main cities are Madrid, the capital, and Barcelona, the leading seaport and industrial center. Most people live in cities and are Roman Catholic, but a large number of Muslims have immigrated.

Portugal

Portugal is a small, democratic country located west of Spain. Most of the land, a low coastal plain, is farmed. Grapes are grown to make wine, and oak trees provide cork. Most Portuguese live in small villages on the coast near Lisbon, the capital, and Porto. Many people fish for a living. Nearness to the Atlantic Ocean helped Portugal become a powerful empire during the 1500s. Today, the European Union helps Portugal's economy by providing **subsidies,** or special payments, to support manufacturing and service industries.

Italy (page 341)

Explaining

Explain why northern Italy is richer than southern Italy.

_____
_____
_____

Describing

What and where is Vatican City?

_____
_____
_____
_____

Italy is a boot-shaped peninsula that extends into the Mediterranean Sea. The Alps dominate northern Italy. The Apennine Mountains run through the central and southern areas, including the Italian island of Sicily. Volcanoes are found throughout the country.

Italy has an industrial economy. Most industry is based in northern Italy in the cities of Milan, Turin, and Genoa. Workers there produce cars, technical instruments, appliances, and clothing. The Po River valley in the north also is a rich farming region where farmers raise livestock and grow grapes, olives, and other crops.

The mountainous land of southern Italy is not good for farming and is poorer than northern Italy. Many people have left the region because it has less industry and higher unemployment.

Italy is a democratic republic. Its people speak Italian, and nearly all are Roman Catholic. About 90 percent live in urban areas. Rome, the capital and largest city, once was the center of the Roman Empire. The Roman Catholic Church is based in Rome. The Church rules Vatican City, which is actually an independent country within Rome's boundaries.

 | **Read to Learn**

Greece (page 342)

Listing

List five components of Greece's economy.

1. _____
2. _____
3. _____
4. _____
5. _____

Identifying

What is the Parthenon?

_____

Greece lies east of Italy and extends from the Balkan Peninsula into the Mediterranean Sea. In addition to its mainland, Greece also has about 2,000 islands. Mountains and poor, stony soil cover Greece, so agriculture is not as important to the economy. People do raise sheep and goats in the highlands, however, and farmers grow wheat and olives in valleys and plains.

The textile, footwear, and chemical industries have become important to Greece's economy in the last few decades. Shipping and tourism also are vital to the country's economy. Greece boasts a large shipping fleet, including oil tankers, cargo ships, and passenger ships. Millions of visitors come each year to see historic sites like the Parthenon, an ancient temple in the city of Athens.

Almost 60 percent of the people of Greece live in urban areas, with one-third of them living in or near Athens, the capital. The people speak Greek, and most follow the Greek Orthodox Christian religion. Greece is a member of the European Union, and it has a democratic republic form of government.

Section Wrap-Up Answer these questions to check your understanding of the entire section.

1. **Explaining** Why has membership in the European Union been good for Portugal?

   _____
   _____

2. **Sequencing** List southern Europe's peninsulas in order from west to east.

   _____
   _____

 *Imagine that you are a reporter assigned to interview a person who has just returned from a trip to southern Europe. Create an outline of important questions to ask in your interview.*

_____
_____
_____

Chapter 12, Section 3

Chapter 12, Section 4 (Pages 348–356)
Eastern Europe

Big Idea

Geography is used to interpret the past, understand the present, and plan for the future. As you read, complete the chart below. Summarize key facts about each group of countries.

Country Groups	Key Facts
Poland, Belarus, Baltic Republics	
Czech Republic, Slovakia, Hungary	
Countries of Southeastern Europe	

Notes — Read to Learn

Poland, Belarus, and the Baltic Republics (pages 349–350)

Identifying

Complete these sentences with the names of the countries they describe.

A flat landscape made _____ easy to invade.

_____ has a command economy.

Poland borders the Baltic Sea. Fertile lowland plains cover most of the country. The Carpathian Mountains rise in the south and west. Throughout history, Poland's flat landscape made it easy for other countries to invade. In 1939 Germany attacked Poland, starting World War II. After the war, a Communist government ruled Poland and set up a **command economy**—the government decided what, how, and for whom goods would be made. Factories made military goods instead of food, which led to shortages. Polish workers and farmers formed Solidarity, a labor group that pushed for democracy and a better life. Communist leaders allowed free elections in 1989, and Poland became a democracy. This event led to the fall of other Communist governments in Eastern Europe. Today Poland has a **market economy**—people and businesses decide what to produce. Agriculture is still important, but industries are growing. Warsaw is the capital.

Belarus, east of Poland, also is covered by a lowland plain. A former Soviet republic, Belarus now has a rigid government and a command economy. Its main resource is **potash,** a mineral used to make fertilizer. Government-controlled farms grow

Notes | Read to Learn

Poland, Belarus, and the Baltic Republics (continued)

Naming

The Baltic Republics are _____, _____, and _____.

grains, and factories make trucks, radios, TVs, and bicycles. Most people are Eastern Orthodox Slavs.

The Baltic Republics—Lithuania, Latvia, and Estonia—were under Russian control until the Soviet Union broke apart in 1991. Now they are democracies with strong economies based on dairy farming, beef production, fishing, and shipbuilding.

The Czech Republic, Slovakia, and Hungary (pages 352–353)

Listing

List six items produced by the Czechs.

_____
_____
_____
_____
_____
_____

Specifying

From whom did the Hungarians descend?

_____

The Czech Republic, Slovakia, and Hungary are landlocked countries. The Czech Republic and Slovakia were one country, Czechoslovakia, from 1918 to 1993. Once under Communist rule, all three countries now have democratic governments.

The Czech Republic features rolling hills, lowlands, and plains bordered by mountains. Prague, the capital, is known for its historic buildings. Czechs enjoy a relatively high standard of living. The country is a major agricultural producer, but manufacturing is the core of its economy. Goods include machinery, vehicles, metals, and textiles, as well as fine crystal and beer.

The Carpathian Mountains tower over northern Slovakia with rugged peaks, thick forests, and lakes. Vineyards and farms dot the south's fertile lowlands. Slovakia has been slow to move to a market economy. It has less industry than the Czech Republic. Most Slovaks are devout Catholics. Bratislava, the capital, is located on the Danube River.

Hungary's landscape is a lowland area with fertile farmland. Its capital, Budapest, straddles the Danube River. Hungary has few natural resources, but it imports raw materials for industry. The country exports chemicals, food, and other products. The Hungarians have a unique language and are descended from the Magyars of Central Asia.

Countries of Southeastern Europe (pages 353–356)

Stating

What physical feature divides Ukraine?

Eleven countries in southeastern Europe are located along the Black Sea or Balkan Peninsula. Ukraine, the largest country in Europe, is located on a lowland plain with the Carpathian Mountains to its southwest. The Dnieper River divides the country into two sections. The lowland steppes to the west have rich soil good for farming and raising cattle and sheep. People of Ukrainian descent live in this "breadbasket of Europe," and they

Chapter 12, Section 4

 **Notes** | **Read to Learn**

Countries of Southeastern Europe (continued)

Summarizing

List two facts about western Ukraine and two facts about eastern Ukraine.

Western Ukraine:

_____

_____

Eastern Ukraine:

_____

_____

Identifying

What are Bulgaria's major economic activities?

_____

_____

want to join the European Union. Coal and iron ore are mined in the eastern plains, which are heavily industrialized. The people in this section are ethnic Russians who want closer ties to Russia. Ethnic divisions have grown sharp.

Romania has coal, petroleum, and natural gas, which helps its economy. Romans once ruled this region, so the Romanian language is based on the Latin spoken in ancient Rome. Bucharest, the capital, is the country's major commercial center.

Moldova is a landlocked country between Ukraine and Romania. It has fertile soil and productive farms but few mineral resources or industries. It is Europe's poorest country.

Bulgaria is a mountainous country with fertile river valleys that are good for farming. Many people work in factories located in Sofia, the capital, or in the tourism industry at Black Sea resorts.

Most countries on the Balkan Peninsula once were part of a Communist country called Yugoslavia. In the early 1990s, Slovenia, Croatia, Bosnia and Herzegovina, and Macedonia declared their independence. Serbia wanted to keep Yugoslavia together under Serbian rule, however. Serbs carried out **ethnic cleansing,** removing or killing entire ethnic groups. The conflicts left the Balkan countries scarred and with even poorer economies than they had under Communist rule. Their mountainous landscape is not good for farming, and they have few natural resources.

Albania is unique in that it is the only European country with a majority Muslim population. Albanian farmers outnumber factory workers, and this country too is poor.

Section Wrap-Up

Answer these questions to check your understanding of the entire section.

1. **Explaining** What was Solidarity, and what was its goal?

   _____

   _____

2. **Sequencing** What caused Serbia to implement ethnic cleansing?

   _____

   _____

 **Expository Writing**

On a separate sheet of paper, discuss the differences between a command economy and a market economy. Explain why you think countries like Poland wanted to move to a market economy.

Chapter 13, Section 1 (Pages 372–375)

Physical Features

Big Idea

Changes occur in the use and importance of natural resources. As you read, complete the diagram below. List six of Russia's major landforms.

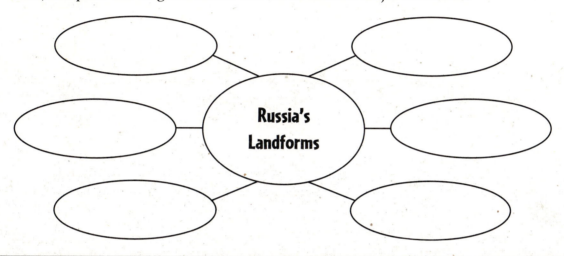

 **Read to Learn**

Landforms of Russia (pages 373–374)

Naming

Name the two continents that Russia straddles and the mountain range that divides them.

Continents:
_____
_____

Mountain range:
_____

Russia is the largest country in the world, measuring about 6,200 miles from east to west. The country is located on two continents—Europe and Asia. The Ural Mountains divide the European and Asian parts of Russia.

Because the country is so large, Russia has a long coastline. Russia is located in the north, however, and the water along most of its coast is frozen for much of the year. Few ports in Russia are always free of ice. In the southwest, though, Russia does have port cities on the Black Sea. The Black Sea provides a warm-water route for ships traveling from inland Russia to the Mediterranean Sea.

European Russia

A variety of landforms cover Russia. Most of the European portion of Russia lies on the Northern European Plain. This fertile area has a mild climate, and about 75 percent of Russia's population lives there. Moscow, the capital, and St. Petersburg, a large port city near the Baltic Sea, are located on the Northern European Plain. Good farmland and grassy plains are found farther south along the Volga and other rivers.

Chapter 13, Section 1

Landforms of Russia (continued)

Describing

Describe the physical characteristics of the three sections of Siberia.

North

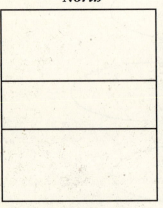

South

Specifying

Why is the Caspian Sea important to Russia?

_____
_____
_____

The rugged Caucasus Mountains rise to the far south in European Russia. Because the mountains are located along a fault line, destructive earthquakes strike the Caucasus area.

Asian Russia

Asian Russia is on the eastern side of the low, eroded Ural Mountains. The region known as Siberia makes up a huge part of Asian Russia. Northern Siberia is a large treeless plain that is frozen most of the year. Few people live in Siberia. Those who do work by fishing, hunting seals and walruses, and herding reindeer. Just south of the plains is a region of dense forests, where people make a living by hunting and lumbering. The southern part of Siberia is covered by plains, plateaus, and mountain ranges.

In the far east, the Kamchatka Peninsula juts into the Pacific Ocean. It is covered by mountains that are part of the Ring of Fire —the rim of the Pacific Ocean where tectonic plates meet. The Earth's crust is unstable along the Ring of Fire, which results in earthquakes and active volcanoes in Kamchatka.

Inland Waters

Russia has many inland waterways. The major river in European Russia is the Volga. In Siberia, many rivers begin in the southern mountains and flow north to the Arctic Ocean. The Caspian Sea, located in southwestern Russia, is the largest inland body of water in the world. A saltwater lake, it is important for its fish as well as its oil and natural gas deposits. Lake Baikal, the world's deepest freshwater lake, is located in southern Siberia. It holds one-fifth of the world's unfrozen freshwater. Baikal, or nerpa, seals and many other forms of aquatic life are found in the deep waters of Lake Baikal.

Natural Resources (page 375)

Identifying

Circle seven natural resources found in Russia.

Russia is filled with natural resources. The country has large reserves of **fossil fuels**—oil, natural gas, and coal. It also has major deposits of iron ore, copper, and gold. The iron ore has been used to build up Russia's steel industry.

Another valuable resource found in Russia is timber. Much of Siberia is covered by trees. Russia provides about 20 percent of the world's softwood. **Softwood,** the wood from evergreen trees, is used in buildings and for making furniture.

Many of Russia's resources are located in the frozen region of Siberia. The immense size and remote location of Siberia

Notes | Read to Learn

Natural Resources (continued)

Summarizing

What challenges does Russia face in accessing its resources?

_____
_____
_____

provide challenges to Russians who attempt to access the region's resources. Siberia lacks the **infrastructure,** or roads and railroads, needed to transport materials. The intense cold makes it difficult for workers to stay warm and to keep their equipment from freezing. Recently, a pipeline was built to carry natural gas from Siberia to Europe.

Section Wrap-Up

Answer these questions to check your understanding of the entire section.

1. **Explaining** Why does Russia not benefit from having a long coastline?

   _____
   _____

2. **Making Generalizations** Where do most Russians live? Why?

   _____
   _____

In the space provided, write a paragraph describing which region of Russia would be the most difficult to live in based upon physical characteristics.

_____
_____
_____
_____
_____
_____
_____

Chapter 13, Section 1

Chapter 13, Section 2 (Pages 378–382)
Climate and the Environment

Big Idea

People's actions change the physical environment. As you read, complete the diagram below. List four factors that lead to Russia's cold climate, especially those related to location and landforms.

```
[  ]        [  ]
   [  ]  [  ]
   [ Russia's Climate ]
```

Notes — Read to Learn

A Cold Climate (pages 379–380)

Listing

List three climate zones that are found in Russia.

1. _____
2. _____
3. _____

Most of Russia is located in the high latitudes. Because it is so far north, Russia does not receive much of the sun's heat, even during the summer. Much of Russia also is inland, away from the oceans. In other parts of the world, the warm currents of the Atlantic Ocean and Pacific Ocean help moderate the temperature. Russia does not have this benefit.

The landforms in Russia also affect the country's climate. In the north, the elevation of the land is not high enough to block the cold, icy air that blows south from the Arctic region. In the south and east, tall mountains block the warm air that would otherwise come from the lower latitudes. As a result, most of Russia experiences only two seasons—long winters and short summers. Spring and autumn are short periods of changing weather.

Most of western Russia has a humid continental climate, with warm, rainy summers and cold, snowy winters. The average July temperature in Moscow is just 66°F. The average January temperature, in contrast, is 16°F. These cold winters have played an important role in Russia's history. Germany's advance into Russia during World War II was stopped by the bitter cold.

A Cold Climate (continued)

Speculating

What is a coniferous tree?

_____

In the northern and eastern parts of Russia, the summers are short and cool, and the winters are long and snowy. In the far north, the tundra climate zone has resulted in a permanently frozen layer of soil beneath the surface, called **permafrost.** Only mosses, lichens, and small shrubs survive in the tundra. The subarctic climate zone is located south of the tundra. It has slightly warmer temperatures than the tundra zone does. The **taiga**—the world's largest coniferous forest—stretches about 4,000 miles across Russia's subarctic region. The taiga is about the size of the United States.

Russia's Environment (pages 380–382)

Finding the Main Idea

What is the main idea of this subsection?

_____
_____
_____
_____

Paraphrasing

Complete this sentence.

Other countries are providing Russia with _____ *to improve* _____ *and clean up* _____ .

Throughout the 1900s, Russia's leaders focused on expanding the country's economy. In doing so, they paid no attention to how this growth damaged the environment.

Factories continue to release pollutants into the air today. **Pollutants** are chemicals and smoke particles that cause pollution. A thick haze of fog and chemicals, called **smog,** hovers over many of Russia's cities. As a result, a large number of Russians have lung diseases and cancer.

Water Pollution

Russia also has water pollution. Agriculture and industry use many chemicals. These chemicals often drain into waterways. Poor sewer systems are another source of water pollution.

One effect of water pollution is that some of Russia's animal species are threatened. For example, animal populations around Lake Baikal may be getting smaller because pollution has damaged the water. Water pollution also affects people. More than half of the people in Russia do not have safe drinking water.

Cleaning Up

Russia is making efforts to clean up its environment. Other countries are providing aid to Russia. Russia is using this assistance to improve the country's sewage systems. International aid also is helping Russia clean up heavily polluted sites.

Cities are building power plants that are more efficient. These new plants will use less energy than the existing ones. They also will burn fuel more cleanly. As a result, fewer pollutants will be released into the air. Some of these efforts are cleaning up the existing pollution, and others are intended to reduce future pollution. Even so, it will take a long time for Russia to have a healthy environment.

Section Wrap-Up

Answer these questions to check your understanding of the entire section.

1. **Determining Cause and Effect** What effects do air and water pollution have on people and animal life in Russia?

   _____
   _____

2. **Specifying** What are three steps Russia is taking to improve its environment? Which steps address existing pollution and which reduce future pollution?

   _____
   _____
   _____
   _____

Expository Writing

In the space provided, write a paragraph providing several suggestions on how a country can expand its economy without harming the environment.

_____
_____
_____
_____
_____
_____
_____
_____
_____
_____
_____
_____

Chapter 14, Section 1 (Pages 388–394)
History and Governments

Big Idea

The characteristics and movement of people impact physical and human systems. As you read, make an outline of the section using the model below. Use Roman numerals to number the main headings. Use capital letters to list two key facts below each main heading.

I. First Main Heading
   A. Key Fact 1
   B. Key Fact 2
II. Second Main Heading
   A. Key Fact 1
   B. Key Fact 2

Notes | Read to Learn

The Russian Empire (pages 389–391)

Sequencing

Write down important dates and events in Russia's development.

_____
_____
_____
_____
_____
_____

Russia began as a small trade center. Slavic people lived along the rivers in Ukraine and Russia. In the A.D. 800s, Slavs settled the town of Kiev, which became the civilization known as Kievan Rus. **Missionaries,** or people who move to another area to spread their religion, brought Eastern Orthodox Christianity and a written language to Kievan Rus in A.D. 988.

In the 1200s, Mongol warriors from Central Asia conquered Kievan Rus. Many of the Slavic people moved north. They built a small trade center called Moscow. It became the center of a new Slavic territory called Muscovy. Ivan III, a prince of Muscovy, declared independence from Mongol rule. A strong ruler, he was known as "Ivan the Great."

In 1547 Ivan IV declared himself **czar,** or emperor, of Muscovy, which became Russia. Known as "Ivan the Terrible," he expanded his empire by conquering neighboring lands. Later czars, such as Peter the Great and Catherine the Great, also expanded the empire. They wanted a warm-water port for trade. They also wanted to become more European. Peter the Great built a new capital—St. Petersburg—close to Europe. The Russian Empire extended to the Pacific Ocean and Central Asia.

Chapter 14, Section 1

The Russian Empire (continued)

Explaining

How did Russia's physical geography act as a weapon in 1812?

_____
_____
_____
_____

The czars, large landowners, and wealthy merchants lived comfortably. The majority of Russians were serfs, however. **Serfs,** or peasant farm laborers, could be bought and sold with the land.

Russia's cold climate and huge size helped defeat invaders. The French emperor Napoleon invaded in 1812, but he lost most of his army retreating during the brutal Russian winter.

Russia changed greatly in the late 1800s. Czar Alexander II freed the serfs in 1861. He built industries and railroads to modernize Russia's economy. Yet most Russians remained poor, and unrest spread. In early 1917, the people revolted and overthrew Czar Nicholas II. Vladimir Lenin established a **Communist state** in which the government controlled the economy and society.

The Rise and Fall of Communism (pages 391–394)

Specifying

Underline the reason Lenin ended private ownership.

Circle the phrase that describes why factory goods were of poor quality.

Defining

What was the Cold War?

_____
_____
_____
_____

Lenin created a new country called the Union of Soviet Socialist Republics (U.S.S.R.), or the Soviet Union. It included 15 republics and many different ethnic groups. Lenin followed the ideas of Karl Marx, a German political thinker. Marx believed that industrialization was unfair because factory owners had much power while the workers had little. Lenin wanted to make everyone equal. He ended private ownership. The government took control of all factories and farms.

Later Soviet leaders, such as dictator Joseph Stalin, set up a command economy in which the government made all economic decisions. The government introduced **collectivization,** a system in which small farms were combined into larger ones. The collective farms were not efficient, however, and did not produce enough food for all the people. Soviet factories produced steel, machines, and military equipment. Without competition, however, many goods were of poor quality.

During World War II, Germany invaded the Soviet Union. About 20 to 30 million Russian soldiers and civilians died. After the war, Stalin kept troops in neighboring countries to make sure the U.S.S.R. would not be invaded again. He set up Communist governments in Eastern Europe.

Although the Soviet Union and the United States were allies during World War II, they became enemies after the war. Because no physical combat occurred, their conflict became known as the **Cold War.** The United States led democracies in Western Europe, and the Soviet Union led Communist Eastern Europe. Many Soviet resources were used to produce weapons.

Chapter 14, Section 1

The Rise and Fall of Communism (continued)

Identifying

Who was the last president of the Soviet Union?

_____

Who were the first two presidents of an independent Russia?

_____

_____

Mikhail Gorbachev became the Soviet leader in 1985. The people were tired of enduring shortages of food and other goods. Gorbachev established the policy of **glasnost,** or "openness." People could say or write their opinions without being punished. He also introduced **perestroika,** or "rebuilding," to boost the economy. Perestroika allowed for small, privately owned businesses. Gorbachev thought these new policies would strengthen the people's support of the government. Instead, Eastern Europeans began to doubt communism, and protests arose. By 1991, all of Eastern Europe's Communist governments had changed to democracies.

Hard-liners in the Soviet Union wanted to stop the changes and return to communism. Boris Yeltsin became president of Russia, the largest of the Soviet republics. In 1991 hard-liners attempted a **coup** to overthrow him by military force. The coup failed. Russia and all the other Soviet republics declared independence. By the end of 1991, the Soviet Union no longer existed as a nation.

Yeltsin worked to build a democracy and to create a market economy. But Vladimir Putin, who became president of Russia after Yeltsin, increased government controls to deal with rising crime and violence. Ethnic minorities, such as those in the Chechnya region, have tried to separate from Russia.

Section Wrap-Up *Answer these questions to check your understanding of the entire section.*

1. **Determining Cause and Effect** What effect did the command economy have on the Soviet Union, particularly during the Cold War?

_____

_____

2. **Explaining** What impact did Gorbachev's policies have on Eastern Europe?

_____

_____

On a separate sheet of paper, write a summary of how Russia started as a small trade center and grew to become the Russian Empire.

Chapter 14, Section 1

Chapter 14, Section 2 (Pages 396–400)
Cultures and Lifestyles

Big Idea

Culture groups shape human systems. As you read, complete the diagram below by listing six details about the arts in Russia.

 **Read to Learn**

Russia's Cultures (pages 397–398)

Identifying

What are the three largest ethnic groups in Russia?

_____

_____

_____

Russia is home to dozens of different ethnic groups. The largest ethnic group includes Russians, or Slavs who descended from the people of Muscovy. The next-largest groups are Tatars, or Muslim descendants of Mongols, and Ukrainians, whose ancestors were the Slavs who settled around Kiev. Many of these groups speak their own language. Most people also speak Russian, the country's official language.

The Russian people were not allowed to practice religion during Communist rule, but now the people have religious freedom. The country's major religion is Eastern Orthodox Christianity. Many Muslims live in the Caucasus region.

The Arts

The arts have always been a central part of Russia's culture. The people had a strong **oral tradition**, meaning they passed on stories by word of mouth. Writers and musicians used these stories or folk music to create new works. One theme in Russian artistic works is **nationalism**, or feelings of loyalty toward the country.

Russia's Cultures (continued)

Naming

Who are two famous Russian writers?

_____

_____

Russia is a center of music and dance. Peter Ilich Tchaikovsky wrote the famous ballets *Swan Lake* and *The Nutcracker*. Two world-famous ballet companies are the Bolshoi in Moscow and the Kirov in St. Petersburg. Russia is also known for its literature, featuring such writers as Leo Tolstoy during the 1800s and Alexander Solzhenitsyn in the Communist era. The Hermitage Museum in St. Petersburg holds many masterpieces. There you can see the czars' jewel-encrusted Easter eggs made by Peter Carl Fabergé.

Specifying

What has been the major area of focus for Russian scientists?

_____

Scientific Advances

Because of its emphasis on science education, Russia has many scientists, mathematicians, and doctors. Some of the scientists' most significant work has been in the area of space exploration. In 1961 Russian Yuri Gagarin became the first person to fly in space.

Life in Russia (pages 399–400)

Listing

What are three popular sports and three national holidays in Russia?

Sports

_____

_____

_____

Holidays

_____

_____

_____

Russia is modernizing after decades of Soviet control. However, it also faces several challenges. Most of Russia's cities are west of the Ural Mountains. The majority of people in these cities live in large apartment buildings. Housing is scarce and expensive, so grandparents, parents, and children often live together. Wealthier people have country homes, called dachas. It is common for people to have vegetable gardens at their dachas. They either eat the vegetables or sell them in the cities.

In many areas of Russia, homes are designed to protect against the extreme cold. In Siberia, for example, some houses have three doors at the entrance. This keeps cold air from coming in when the outside door is opened.

Sports and Holidays

The most popular sports in Russia are winter sports or sports that are played indoors. Russian athletes are among the world's best in hockey, figure skating, and gymnastics.

Russians celebrate several national holidays. The newest holiday is Independence Day. It occurs on June 12, and it celebrates Russia's declaration of **autonomy**, or independence, from the Soviet Union. Another festive holiday is New Year's Eve. In the spring, there is a week-long holiday called *Maslenitsa* to mark the end of winter.

Chapter 14, Section 2

Life in Russia (continued)

Explaining

What are two challenges that Russia faces?

1. _____
   _____
2. _____
   _____

Transportation and Communications

Railroads are the major way of moving people and goods throughout the vast land of Russia. A network of railroads covers the heavily populated area west of the Ural Mountains. This network connects to the Trans-Siberian Railroad. Completed in the early 1900s, the Trans-Siberian Railroad is the longest rail line in the world. It runs from Moscow in the west to Vladivostok in the east. The railroad makes it possible for Russians to use Siberia's natural resources.

Russia does not have a good highway system. No multilane highways connect cities. The roads that do exist are in poor condition. Car ownership is on the rise in Russia, so more and better roads are needed. The government is currently building a 6,600-mile national highway across the country.

Russia is also working on improving its communications systems. Upgrades have been made in the telephone system, including installing new phone lines for faster transfer of Internet information. Many rural areas still have poor phone service, however.

Section Wrap-Up

Answer these questions to check your understanding of the entire section.

1. **Describing** What effect does the scarcity and expense of housing have on Russian lifestyles?

   _____
   _____

2. **Explaining** What is the government of Russia doing to solve some of the country's transportation problems?

   _____
   _____

On a separate sheet of paper, write an oral tradition that has been passed down in your family, including how long the story has been passed down.

Chapter 15, Section 1 (Pages 408–412)
A Changing Russia

Big Idea

Geographers organize the Earth into regions that share common characteristics. As you read, complete the diagram below. Describe three major effects of the fall of communism on Russia.

Notes — Read to Learn

Changing Politics and Society (pages 409–411)

Specifying
List three aspects of the new Russian government that were determined by Russian voters.

1. _____
_____
2. _____
_____
3. _____
_____

The fall of communism in 1991 brought changes to Russia's government, economy, and society.

A New Form of Government

After communism fell, Russia became more democratic. In 1993 Russians voted on a new constitution, elected representatives to the legislature, and elected their first president—Boris Yeltsin. Today the Russian Federation is a federal republic, with power divided between national and regional governments.

A New Economic System

Russia has tried to move from a command economy to a market economy. One feature of this new economy is privatization. The goal of **privatization** is to shift the ownership of businesses from the government to individuals. Businesses now have to compete with one another to produce goods that Russian consumers need and want—and at a price they are willing to pay.

Changing Politics and Society (continued)

Summarizing

Underline three positive effects of Russia's new political freedom.

Circle three negative effects of Russia's new economic freedom.

Changes in Society

Russians now have political freedom. They may join different political parties and are allowed to criticize leaders and their policies. The government no longer controls news reports and books. Russians have more contact with American and European ideas, music, and fashion. Consumerism, or the desire to buy goods, has led to the emergence of a middle class in Russia. People in the **middle class** are neither rich nor poor. But they can buy cars, electronics, and new clothing.

New economic freedom did not guarantee success, however. Some businesses failed. Some Russians lost their jobs. Other workers face **underemployment,** meaning they work at jobs for which they are overqualified. Many people must have two jobs to survive. Russia also has many **pensioners** who are unable to work and receive a fixed income from the government. In a market economy, prices go up and down. Pensioners' incomes do not go up, so pensioners cannot always afford to buy goods.

Population Changes

During Soviet times, many ethnic Russians moved to other parts of the Soviet Union. When those republics became independent, ethnic Russians were no longer welcome. Many decided to return to Russia. Even so, Russia's population has declined. Low birthrates and rising death rates are the cause.

Russia's Economic Regions (pages 411–412)

Identifying

Identify Russia's four economic regions.

1. _____
2. _____
3. _____
4. _____

The Moscow Region

The city of Moscow is the political, economic, and transportation center of Russia. The Moscow region is home to much of Russia's manufacturing. During Soviet rule, factories focused on **heavy industry,** or producing goods such as machinery, mining equipment, and steel. Today many factories have changed to **light industry,** or producing consumer goods such as clothing and household products.

St. Petersburg and the Baltic Region

St. Petersburg and the Baltic region are located in northwestern Russia, near the Baltic Sea. St. Petersburg is a cultural center, attracting thousands of tourists. The city also is a major port, making it an important trading center. Factories there make ships, machinery, and vehicles. The people of St. Petersburg buy their food and fuel from other regions in Russia.

Russia's Economic Regions (continued)

Paraphrasing

Complete the following sentences.

The Volga River is in the _____ and _____ region.

The river provides water for _____ and for _____ farms.

Another Russian port city on the Baltic Sea is Kaliningrad. It is located on a small piece of land between Poland and Lithuania, isolated from the rest of Russia. Goods shipped to Kaliningrad from other parts of Russia actually have to cross through other countries to get to this port. Kaliningrad is Russia's only port on the Baltic Sea that stays ice-free all year.

The Volga and Urals Region

To the south and east of Moscow is the Volga and Urals region. This region is valued for its manufacturing and farming. The Volga River is a vital waterway used for shipping people and goods, and it carries nearly half of Russia's river traffic. The Volga River also provides water needed for hydroelectric power and for irrigating farms. Wheat, sugar beets, and other crops are grown in this region. Many of Russia's natural resources come from the Ural Mountains. Minerals such as copper, gold, lead, nickel, and bauxite are found there.

Siberia

Russia's fourth economic region is Siberia. This region holds valuable iron ore, uranium, gold, and coal. Timber is harvested from the taiga. Cold Arctic winds and frozen ground make it difficult for Russia to take advantage of Siberia's resources, however.

Section Wrap-Up

Answer these questions to check your understanding of the entire section.

1. **Comparing and Contrasting** Compare and contrast the Moscow region and the St. Petersburg/Baltic region.

   _____
   _____

2. **Analyzing** What barriers prevent making use of Siberia's natural resources?

   _____
   _____

On a separate sheet of paper, write an editorial either for or against Russia's move from a command economy to a market economy. Give reasons to support your opinion.

Chapter 15, Section 1

Chapter 15, Section 2 (Pages 418–422)
Issues and Challenges

Big Idea

Geography is used to interpret the past, understand the present, and plan for the future. As you read, complete the diagram below. Identify six changes, both positive and negative, that have resulted from Russia's switch to a free market economy.

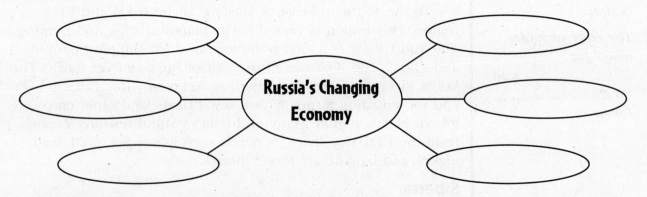

Read to Learn

Political and Economic Challenges (pages 419–421)

Speculating

What is a decree? In your own words, explain why decrees are undemocratic.

_____
_____
_____
_____
_____

Roadblocks to Democracy

Becoming a democracy has not been easy for Russia's people. Confusion over government powers is a problem. In the new government, the Russian president has the power to issue **decrees,** or rulings that have the force of law but do not need the legislature's approval. This power gives the president much control over the country.

Russia is a federal republic, which means that power is shared among national, regional, and local governments. To ensure that leaders obey his wishes, however, President Vladimir Putin organized the country into seven large districts. He then appointed his supporters as district governors.

Many Russian politicians ignore democratic practices. The courts and legal system tend to favor wealthy citizens. Most of the people do not understand how the government works, which prevents them from being able to change it.

Political and Economic Challenges (continued)

Explaining

Explain why banks are important in an economy.

_____
_____
_____
_____
_____
_____
_____

Identifying

In which region of Russia has a violent separatist movement occurred?

_____

Shifting to a Market Economy

In Russia's market economy, new companies have been started, some incomes have risen, and higher prices for oil and gas exports have brought more money into the country. But oligarchs control various parts of the economy. **Oligarchs** are members of a small group of rulers that holds much power. Russia's oligarchs are often corrupt business leaders.

The benefits of economic success have not reached all the Russian people. Some Russians have become wealthy, but others have become even poorer. Economic success varies widely from region to region, too. Incomes in Moscow are much higher than incomes in other Russian cities.

Russia's banking system has hindered economic growth. Banks play an important role in an economy. People deposit their savings into banks to earn interest. Banks, in turn, loan that money to other people who buy homes or cars, or start new businesses. These actions create jobs. In Russia, however, many people do not trust the banks. They are afraid that they will lose the money they deposit. To remedy this problem, the government created a **deposit insurance** system that promises to repay people who deposit money in a bank if that bank goes out of business.

Challenges to National Unity

Regional rivalries have increased in recent years and make it difficult to unify Russia. When the Soviet Union fell apart, some of the ethnic groups in Russia wanted to form their own countries. This gave rise to **separatist movements,** or efforts to break away from the national government and become independent. A violent separatist movement began in Chechnya, a region near the Caucasus Mountains in southern Russia. Russia's then-President Boris Yeltsin gave the region more self-rule, but many of the Chechen people wanted complete independence. In 1994 the Russian army was sent to Chechnya, and both sides suffered heavy losses. The situation remains unresolved.

Chapter 15, Section 2

Russia and the World (page 422)

Finding the Main Idea

What is the main idea of this subsection?

_____
_____
_____

Russia is a major world power and is prominent in world affairs. The country has worked in recent years to help strengthen its connections with other countries. For example, in 2002 Russia supported the United States and other North Atlantic Treaty Organization (NATO) countries in their fight against global terrorism.

At the same time, many countries, including the United States, are troubled by President Vladimir Putin's increased power. They are concerned by Putin's moves away from democratic practices. Tension also is increasing between Russia and former Soviet countries. Some Russian leaders want to have more influence in these countries. Although neighboring countries are unhappy with Russia's actions, they depend on Russia for oil and natural gas.

Section Wrap-Up

Answer these questions to check your understanding of the entire section.

1. **Describing** What challenges face the Russian people as they shift toward democracy?

   _____
   _____

2. **Explaining** How have regional rivalries hurt Russia?

   _____
   _____

In the space provided, write a paragraph describing Russia's role as a world power and predicting the role Russia will play in world affairs over the next several years.

_____
_____
_____
_____
_____